On the Origin of Form

On the Origin of Form

Evolution by Self-Organization

Stuart Pivar

Foreword by
Mark A. S. McMenamin, PhD

North Atlantic Books
Berkeley, California

Published by
North Atlantic Books
P.O. Box 12327
Berkeley, California 94712

Cover art from a seventeenth-century engraving by J. Mynde
Cover and book design by Susan Quasha
Printed in the United States of America

On the Origin of Form is sponsored by the Society for the Study of Native Arts and Sciences, a nonprofit educational corporation whose goals are to develop an educational and cross-cultural perspective linking various scientific, social, and artistic fields; to nurture a holistic view of arts, sciences, humanities, and healing; and to publish and distribute literature on the relationship of mind, body, and nature.

North Atlantic Books' publications are available through most bookstores. For further information, call 800-733-3000 or visit our website at www.northatlanticbooks.com.

Library of Congress Cataloging-in-Publication Data

Pivar, Stuart, 1930-
On the origin of form : evolution by self-organization / Stuart Pivar ;
foreword by Mark A. S. McMenamin.
p. ; cm.
Includes bibliographical references and index.
ISBN 978-1-55643-886-8
1. Morphogenesis. 2. Embryology. 3. Evolution. 4. Germ
cells—Ultrastructure. 5. Torus (Geometry) I. Title.
[DNLM: 1. Morphogenesis—genetics. 2. Evolution. 3. Germ Cells. 4.
Models, Biological. QH 491 P693o 2009]
QH491.P58 2009
571'.833—dc22

2009018008

1 2 3 4 5 6 7 8 9 UNITED 15 14 13 12 11 10 09

Contents

List of Plates

"Mysteries are not necessarily miracles."

Johann Wolfgang von Goethe

Author's Note

From One Cell to Complex Organism

Before Darwin, natural history (biology) was divided into two disciplines: morphology (shape) and physiology (function). These were neatly separate until Darwin suggested that one causes the other. This mind-boggling idea has been the basis of the inquiry into the cause, or causes, of evolution, and of vigorous contention among scientists that continues today. The debate is not over whether evolution took place, but over its primary mechanisms.

The arch-morphologist of the enlightened eighteenth century was scientist/poet Johann Wolfgang von Goethe (1749–1832), who coined the term "morphology." Goethe pursued a life-long, Faustian quest for the presumed single, universal structure he called the "Urform," which underlies and shapes all living form. His correspondence indicates on several occasions that he thought he had found the elusive structure.

Two hundred years later the discovery that organic forms could be simulated by the deformation of a strange, newly discovered topological surface led to the conjecture that this unique structure is the basis of all plant and animal shapes, the Urform of Goethe. This structure is an unusual topological configuration consisting of a torus (doughnut form) within a torus.

This discovery is the basis of this book, presenting what evolutionary biologists call a "structuralist" model of biological origins. The model is based on the premise that the species body form is encoded not in the DNA, but in the patterned structure of the membrane of the primordial germ plasm, universal predecessor of the egg. After introductory notes on the scientific phenomena that are touched on by this premise, the theory and the model itself is presented. Explication of the multi-torus model is followed by reprints of scientific papers that form the basis for its premise—the toroidal nature of the embryonic tissue.

Also included in this book is a review of the body of scientific literature which questions today's biology dogma, called the Modern Synthesis or Neo-Darwinism, the dogma installed by committee in the 1940s, claiming that organic shapes are formed by natural selection for adaptation to the environment, based on advantageous mutations caused by random genetic errors.

In the past few years a defining architectural structure that occurs in the embryos of animal phyla has been identified in a series of publications by biologists E. Presnov, V. Isaeva, L. Beloussov, Y. Kraus, E.H. Davidson, H. Jockusch, and A. Dress. These discoveries rewrite the understanding of the embryonic membranes that shape the form of the body.

The torus model shows that there is a plausible alternative to Natural Selection and Intelligent Design, neither of which explain the evolution of complex form. This problem is exacerbated by the use of the term "Darwinism" by some writers who use it interchangeably for both the fact of evolution by modified descent, well known to society in Darwin's time, and natural selection, the mechanism proposed by Darwin to account for it. All scientists believe in evolution. Many believe that natural selection plays a lesser, editorial role in its working mechanism. Paleontologist Stephen Jay Gould called those who champion natural selection and adaptation as the sole mechanism of evolution "Darwin Fundamentalists."

While at this writing a number of scientists of world reputation accept the "torus premise" as plausible, many biologists are unable to accept its plausibility as it presumes the negation of a lifetime belief system. As a prominent American Museum of Natural History scientist has said, "I think it's right, but no one will believe it, including me."

Stuart Pivar
New York City
2009

Road Map

There are nearly a hundred titles currently on the shelves of bookstores dealing with evolution, Darwin, and the origin of life. They discuss as a common theme the theories, ideas, and hypotheses by which the secrets of how evolution works will eventually be solved. Included in these "solutions" are natural selection and the decoding of the DNA molecule.

This book does not fall into that category. It instead proposes a hypothetical solution to the problem of the origin of complex life from a single cell. Graphic "blueprints" are shown which purport to account for the forms of animal and plant life, depicting the path from egg to the complex organic forms.

The role of natural selection in evolution has been debated since Darwin's day, beginning with the great man's declaration in later editions of *On the Origin of Species by Natural Selection* that natural selection is not the sole agent of evolution. The strong consensus of today is that natural selection acts to eliminate less fit species that are formed by other means. The other means are, for the most part, not specified, but are generally referred to by the term "self-organization." The term implies that the body becomes organized by internal effects, much as do atoms, molecules, or crystals.

This book proposes a model of self-organization. The model is based on a discovery published in 2003 in the *Bulletin of Mathematical Biology* by Harald Jockusch and Andreas Dress entitled "From Sphere to Torus, a Topological View of the Metazoan Body Plan." In this short, seven page paper the authors demonstrate that the body plans of complex animal and plant life are generally in the form of a topological surface they have dubbed the "multi-torus." They exemplify the torus form as that of a tire or a bagel. This discovery was the needed corroboration after many years of investigation which led to the verification of this theory. The following is the abstract (opening premise) of their paper [the full paper is reproduced at the end of this book]:

> From the viewpoint of mathematical topology, membrane systems in intact living cells can be described as closed and orientable surfaces, i.e., as surfaces with two sides and no boundary lines so that an 'inside' and an 'outside' can be distinguished. Usually, biomembranes represent topological spheres, often one embedded in another one. Toroidal membranes are occasionally observed,

> e.g., in specialized structures of plant cells like the prolamellar body. Here, we propose that rules analogous to those that govern the topology of biomembranes hold for the epithelial cell sheets that cover anatomically external as well as internal surfaces of multicellular animals. We suggest to study the emergence of morphological complexity during metazoan development using concepts from mathematical topology, and propose experimental analyses of those topological transitions that appear to be relevant in development and evolution.

The author's laboratory, comprised of scientists and scientific illustrators, has demonstrated that the body forms of the animal and plant phyla can be accurately simulated by the deformation of an elongated sphere with an internal canal connecting front and back—a topological surface in the category known as the torus. And, that one torus within another can account for many complex biological forms, from the shape of a fly, a flower, or a human, to the stripes on tigers and zebras, and the patterns on butterfly wings. This discovery took place as investigators fabricated experimental inflatable toroidal balloons and observed that these structures, filled with air or various fluids, instantly generate repeatable surface patterns which simulate diverse organic forms when compressed or bent. The tradition of simulating biological form with inflatables began with the nineteenth-century embryologist Wilhelm His, and continues today with the work of Adolph Seilacher and his followers.

The toroidal nature of the cell membrane has been discussed recently in various publications dealing with the topology of the cell membrane, notably in the work of Eugene Presnov in *Bulletin of Mathematical Biology* (2006) and Yulia Kraus in *The International Journal of Developmental Biology* (2006). In *Science* (2007) J. Smith, C. Theodoris, and E. H. Davidson describe the membrane of the sea urchin embryo as a dynamic torus. Two of these papers are excerpted in Part III.

The predictive success of this model, or algorithm, is demonstrated in the succeeding pages as being beyond mere coincidence, suggesting that nature herself may be using this process. To account for this phenomenon, a theory is proposed, named "the torus theory." The theory states that nature originates and encodes form in the membrane of the biological entity called the primordial germ plasm, the amoeboidal cell that is the universal predecessor of the egg. This membrane is known to be in the form of a torus.

To account for this theory, and reconcile it with modern observation, two phenomena are invoked:

1. The suggested mechanism by which the encoded pattern of the form is transferred and imparted to the embryo is called the morphogenetic field theory, an acknowledged concept in modern biology.
2. The means by which nature has managed to conceal this secret until now lies in the details of the formation of the embryo, described in the embryological phenomenon called recapitulation.

The sequence of the stages we observe of the formation of the complex embryo is like a shuffled deck of cards. The unravelling of this part of the mystery lies in the idea that the stages of the development of the embryo are a replay of the stages of the embryological development of the entire history of the evolution of the species, epitomized in the catchphrase familiar to biologists, "ontogeny recapitulates phylogeny." The theory of recapitulation which originated in the nineteenth century has been under debate ever since. Although rejected under the neo-Darwinism dogma for much of the past century, it has been seriously reexamined by current evolutionary biologists, notably Stephen Jay Gould.

Biologists quickly see that the predicted models of the torus theory do not correspond with the microscopic observation of the well-known steps in the growth and development of the embryo. According to the theory of recapitulation, evolution works by continually adding new stages to the end of the sequence of embryological development as the organism becomes more and more complex. Since the time alloted for embryological development remains short and constant, the adding of new end stages causes the initial stages of the ancestral predecessors of today's species to be compressed and pressed backward in eons of succeeding evolution until the ancient, initial stages disappear entirely, leaving only the final stages for us to observe, much like seeing only the last few frames of a movie. This theory of the toroidal origin of form reconstructs the lost initial stages. This complex subject is elaborated in a later chapter by the reprinting of excerpts from Stephen Jay Gould's landmark book, *Ontogeny and Phylogeny* (1977).

Biology in the 1920s fell into two camps: those who believed that the inheritance of body form lay in the complex structure of the egg cytoplasm, and those who believed it was encoded in the chromosomes. DNA won. In 1953, the discovery of DNA's vast structure led everyone (including myself) to believe that this must indeed be the blueprint of life. But now, over half a century later, no code has been found, and biology is left with no model theory to account for form. This book proposes that form is encoded neither in the chromosomes nor in the cytoplasm, but in the membrane that encloses them. The premise is that body form is encoded in a pattern woven into the texture of the egg cell membrane, akin to the names woven into Madame Defarge's knitting in Charles Dickens' *The Tale of Two Cities.*

The torus theory is based on a series of well-accepted scientific phenomena that have never before been presented in the proposed combination. These principles are summarized below, with the intent of preparing the reader for the later chapter that presents the model in the format of a scientific paper entitled "The Self-Organization of Biological Form."

This book proposes hypothetical answers to the following fundamental questions:

1. How is the complex body formed from a single egg cell?
2. How did the first one-celled organisms originate?
3. How did today's complex living forms evolve from one-celled organisms?

All life begins with the egg. The hypothesis proposes that the resultant body shape of species is determined by an engrained pattern in the membrane that envelops the egg.

The Life Cycle

The background scenario of the working hypothesis of this book is based on the general plan of life first described by August Weismann in the 1880s called the "Life Cycle," and which is held today as the classic understanding of the organization of the sequential stages of the course of the lifetime of the individual organism.

Billions of years ago the primordial ooze produced protoplasm. A speck of protoplasm enveloped itself in a bilayer membrane skin made of even rows and columns of lipid molecules, forming the first cell—a tubular, amoeboid, torus-shaped membrane that streams perpetually out the front, around the back, and into the rear, causing the well-known phenomenon called "pseudopodal locomotion." The amoeboid cell grew, and periodically split in two, repairing itself after each division. This is a likely origin of the one-celled organisms that exist to this day.

Some of the original, ancient cells came to split into two parts, where one of the two continued to subdivide, forming the multi-cellular form called the "somatic body." The other part, called by Weismann the "immortal germ-cell," is trapped within, traversing the mortal somatic body, growing to form the egg, and exiting before the somatic body dies, but not before it splits again and makes a new body, eternally repeating the cycle. The somatic body evolved over time, assuming an ever larger, more complex form, dictated by the inherited information in the immortal, ancient, germ-line cell from which it is formed in each generation, itself never changing. The body is formed from materials produced by the genes, introduced into the germ cell.

The torus theory suggests that the streaming over time may be plausibly predicted to have arranged the molecules of the membrane in an engrained grid pattern. This immortal pattern encodes the form of the body, as the somatic sister cell, composed initially of the same membrane, grows and subdivides, guided as the cells proliferate by the inherited membrane pattern, as well as by the genomic chemicals within an introduced spherical nucleus.

Summary of the Life Cycle

1. The egg cell membrane is derived from its predecessor, the primordial germ plasm.
2. The germ plasm membrane is in the shape of an amoeboid, tubular torus (a sphere with an internal canal connecting opposite poles) that bears an engrained pattern, the result of eons of amoeboid streaming.
3. The egg cell membrane is a "morphogenetic field" which has the ability to reconstitute its shape after being subdivided into separate cells following fertilization.
4. This reassembly guides the formation of the embryo.
5. Different species result from differences in the mechanical deformation of an ancestral toroidal shape caused by differences in the flexibility of the membrane, resulting from the generation of chemicals controlled by the genes.
6. The mechanical deformation of laboratory models of the shape can accurately simulate the complex animal and plant forms of nature.

The Torus Theory Is Comprised of Two Parts

1. The topological algorithmic model possesses the ability to predict form. That the shape of the body of insects, crustaceans, vertebrates, and flowers can all be accurately simulated by simple deformations of a torus-shaped surface is a geometrically, or topologically, demonstrable fact, corroborated by experiments and illustrations, that cannot plausibly be ascribed to coincidence.

2. The theory holds that form has its origin in the patterned membrane of the germ plasm. How then could nature be using this mechanism? The proposed answer lies in the fact that all individuals originate in the primordial germ plasm, a blob of protoplasm enclosed in a torus-shaped membrane, a patterned grid of rows and columns of lipid molecules, constituting a morphogenetic field. The theory holds that following fertilization the egg cell sub-divides, its membrane distributed among the hundreds of separate cells; that these then quickly come together in the universal embryological events called

blastulation and gastrulation, forming the embryo, invisibly guided by their ability to "recognize" their former neighbors on the egg cell membrane, reassemble the pattern encoded in the immortal, primordial germ plasm membrane, conferring the species form to the body, guided by the properties of the morphogenetic field.

This is a causative, mechanically coherent model of the origin of the complex body from a single cell. The model is built on the fact that all bilateral animals share a simple common body form: a segmented tube with paired, jointed appendages—limbs, mouthparts, and wings. The model is described in the many illustrations showing how the gamut of animal body parts may be simulated by the mechanical compression and bending of a lab model of the membrane, beyond the control of the human operator. This graphic model of life origin demonstrates that biological form is not random, but instead, like atoms, molecules, crystals, planets, and the universe, is self-organized, guided by simple, fundamental physical laws.

A valid question is, Why are there no photographic images of the pattern of the germ plasm which are hypothesized as its basis? The pattern presumed to exist on the surface of the germ plasm that can account for the algorithm is the alignment of the lipid molecules comprising its surface. Although it is not inherently impossible to "see" this, it would require sophisticated microscopic technology well beyond the scope and resources of this current investigation. This may well be fruitfully pursued by properly equipped investigators.

Unintended by-products of this premise are two modalities of integrative medicine given scientific corroboration by the torus theory:

1. The practice of acupuncture is based on the theory that the body surface is empatterned with meridian lines that enter the body, pass through the internal organs, and form a series of circuits that connect into one continuous flow. The theory that body form is the result of the enlargement of the meridianal streaming toroidal surface of the germ plasm provides a model that allows for the possibility of a meridianal configuration of the body surface upon which the theory and practice of acupuncture depends.
2. The concept called "bio-energy fields" may have an origin in the morphogenetic fields that encode the body form.

Other Implications

Astrobiology: A non-random model of life form enhances the likelihood that life in space will resemble the terrestial model as do space rocks and minerals, as it does not depend on the unlikely occurrence of the reenactment on another planet of the entire historical sequence of environmental events on earth.

Creationism: Creationists and Intelligent Design proponents correctly state that natural selection does not provide a complete account for evolution, from which they falsely conclude that evolution never happened. The present account can refute this argument.

Nature's Enlarging Machines

The discovery of the geometric origin of form exposes nature's secret process: the body is an enlargement of its own egg.

Cells subdivide in two systems. Most phyla produce an egg which subdivides into a thousand separate cells which form the embryo and then proliferate separately, causing enlargement with deformation of the image. Insects use a different system called the "syncytium." The self-organized, elongated, segmented insect egg contains a hundred genetic nuclei within its cytoplasm. These migrate to the surface as the egg grows and develops. Then each nucleus encloses itself within a membrane finally forming a tissue of separate cells, by which time the egg has begun to form the embryo on its own. Each cell will proliferate, resulting in the enlargement of the embryonic structure, often accompanied by substantial and extreme deformations.

Except in simple animals, proportional variations and deformations in the adult form disguise its simple geometric origin, that of a segmented, elongated toroidal surface, the form of the primordial germ plasm, and also the presumed form of the primordial ancestors of complex life. Biodiversity comprises variations so vast and extreme that they succeed in overwhelming and obscuring the theme itself.

Nature has no goals. Primordial protoplasm-filled lipid bubbles form, grow, and subdivide according to the laws of physics. This bubble geometry-driven process creates enlargements of a fundamental, self-organized structure, as do inorganic crystals. The body form, far from being randomly generated as is widely believed, may be accounted for rationally, as though a living crystal.

From insect embryogenesis it is clear that the process of cellular subdivision is separate from the morphologically formative forces which actually shape the embryo, and serves to obscure the workings of the latter. As with nature's own enlarging method, a twelve-foot model in Paris was used to make the Statue of Liberty.

The Torus in Topology

Topology, like geometry, is a branch of mathematics. Geometry defines shape by position and dimensions; topology by elastic surfaces. Topology recognizes two categories of enclosed surfaces—the sphere and the torus, the latter being a sphere with a hole through it, like a doughnut or an inner tube. Tori may also have multiple holes, the number (n) of which determines the "genus." A sphere is a torus where n is 0.

The discipline of topology does not utilize the term multi-torus. Nor does its literature seem to contain any references to "inscribed" tori, that is, the torus within a torus. The recent publications of Jockusch and Dress and several other investigators establish the "multi-toroidal" nature of the embryonic tissue. That this configuration is a torus within a torus is the hypothesis of this author.

In math parlance, these are inscribed tori, or a nest of tori. The molting of insects results from the periodic generation of new bodies using the inner contour of the existing body as a mold, the latter to be sloughed off intact, burst by the new expanding inner body. Natural form takes its character from the inward directed growth of the body borne imprisoned in a shell.

The peculiar character of inscibed tori is that the inner cable of the outer torus passes through the inner canal of the inner torus, restraining mutual revolution. If mechanically forced to occur, the inner toroidal surface is sliced in two, or else the cable is imbedded deeply in the midline of its back. This model may reconstruct missing early stages in the observed sequences of cell division or vertebrate embryogenesis (see Plate 32).

From Sphere to Torus

The plant and animal body plan is an enlargement of the structure self-generated by the expansion of a self-organized spherical bilayer. The uniform expansion of the surfaces of a spherical bilayer will produce a bilayer torus. The uniform expansion of a mono-layer spherical surface will produce a larger spherical surface, like an inflating balloon. If the surfaces of a bilayer sphere (two contiguous spheres) expand uniformly, the inner sphere will come under compression and expand inwardly. The laws of mechanics predict first that the surface will shrivel in annular rings, and then opposite poles will indent, meeting in the middle, generating thereby an axially segmented bilayer torus. The surface may delaminate producing two interlocking, independently expanding tori. This configuration is the basis of the body form of plants and animals.

It may well be that the primordial seas teemed with bilayer spheres on their way to becoming tori (see Plates 19, 22, 28, 29).

Preface

Mark A. S. McMenamin, PhD

We are entering exciting times in Biology. We might call this the age of post-natural selection evolutionary biology, in other words, a kind of post-modernism for the natural sciences. The Darwinian fixation on natural selection and its consequent pan-selectionism have proved inadequate for the demands placed upon them, and now we must look elsewhere for a fuller understanding of the evolutionary process. A paradigm shift of the first order is in the process, making this an especially good time to review our understanding of the scientific process as we seek a way forward.

It was once said of the scientific method that science "does not admit of any set method, but must be attempted in every way possible." As a way of knowing, science has an almost sacred character and thus deserves our best efforts to bring forth ideas and evaluate them as best we can. Thus, suppression of ideas really has no place in science.

Scientific controversy is an inevitable and even healthy aspect of scientific study. However, when the rancor generated by competing personalities in science gets too intense, results proceeding from the weaker party can be suppressed or even virtually eliminated, at great loss to the conduct of science. On several occasions I have had to rescue important scientific results from obscurity, or even complete elimination, the result of this type of scientific conflict.

A striking example of this involves the work of Corneille-Jean Koene. Koene's atmospheric science work, with its seminal contributions regarding the origin of carbon dioxide in the air, was nearly lost. This was evidently partly due to controversy with his chief rival, Belgian chemist Jean Servais Stas. Surviving copies of Koene's primary works are so few that one wonders if the conflict led to suppression or even destruction of copies of Koene's book. To remedy this situation, I translated his rarest book into English and published a bilingual edition in 2004 (Mellen Press) as "An English Translation of *The Chemical Constitution of the Atmosphere from Earth's Origin to the Present, and Its Implications for Protection of Industry and Ensuring Environmental Quality* by C. J. Koene (1856)." I took this further with the 2007 publication of "Memoirs of Chemistry (1856) by C. J. Koene; a facing-page english translation of the french text *Mémoires de Chimie*." I thus rescued both of these important books from near oblivion, saving them for future generations of science scholars.

Stuart Pivar's book, *On the Origin of Form* (2009), contains ideas that deserve full scientific scrutiny, especially in light of the turmoil roiling evolutionary biology at present. Pivar is presenting, in a series of brilliantly rendered graphical diagrams that show his interpretation of how modifications of a torus shape can generate a vast panoply of biotic form, a new theory of morphogenesis. Some conventionally oriented evolutionary biologists will feel threatened by this new perspective. Genes can no longer be seen as some kind of self-sufficient blueprint for metazoan organization. Rather, morphogenetic field analysis is needed to understand the morphology and ontogeny of a variety of creatures.

Some of the transitional stages shown by Pivar will appear unfamiliar to embryologists, and may thus invite criticism of the model similar to the way that Haeckel has been criticized for his inaccurate drawings of embryos. I urge readers to suspend disbelief on this matter, however, for the purposes of full evaluation and honest scrutiny. We can't afford to wait a century and a half this time, as was the case with Koene's atmospheric science. Each of Pivar's sub-models needs to be evaluated and tested from the perspective of adult morphology, fossil form, embryological change as modified by condensation, self-organization where appropriate, and finally, and all importantly, morphogenetic field analysis.

This is a seismic event for science. Conventional evolutionary biologists are right to be very worried about this, because it has the potential to trigger the complete collapse of Modern Synthesis Biology. Discerning researchers will act now to stay clear of the falling wreckage. New research is urgently needed, and it is for a very good cause as it has the potential to inject life back into biology. For example, an undescribed type of Ediacaran fossil shows a morphology that is astonishingly in accord with the predictions and main tenets of the morphogenetic torus model. When this fossil is fully described and published, it promises to open a new window on how morphogenetic field analysis can help us understand both ontogeny and phylogeny in exciting new ways.

M. A. S. McMenamin, PhD
Paleontologist, Professor of Geology,
Chair of Earth and Environment
Mount Holyoke College

I

Self-Organization and Natural Selection

Introduction

Biodiversity is the study of how species differ, by describing the adult forms of complex animals and plants. But nature may be better understood by the study of what species have in common. The nineteenth-century morphologists sought the ancestor of form in the early stages of simple animals. Ernst Haeckel famously noted that the earlier the stage of the embryo, the more it resembles the embryos of more distantly related species. What then is the absolute zero of embryogenesis, the one form common to all life, the elementary particle of biology? Goethe called it the *Urform*.

This book is about the discovery of the *Urform*. It demonstrates graphically that the gamut of living forms may be accurately simulated by the deformation of a single simple structure, a tubular torus. Since the germ plasm is tubular in shape, as are the egg membrane, the blastula, the embryo, the larva, and the adult, then it is plausible that this may be the key to a solution to the question of morphogenesis and evolution. The proposed mechanism for the transfer of the encoded pattern engrained in the membrane of the germ plasm as a guide to the formation of the embryo is the phenomenon known as the morphogenetic field, to be described in the succeeding pages.

The Germ Plasm Theory

According to the Germ Plasm Theory, postulated by nineteenth-century biologist August Weismann, the germ plasm, which is independent from all other cells of the body called the somatoplasm, is the essential element of germ cells (eggs and sperm) and is the hereditary material that is passed from generation to generation. Weismann first proposed this theory in 1883; it was later published in his treatise Das Keimplasma (1892; The Germ-Plasm: A Theory of Heredity). This view contradicted Lamarck's theory of acquired characteristics, which was a prevalent theory of heredity of the time. Although the details of the germ plasm theory have been modified, its premise of the continuity of hereditary material is the basis of the modern understanding of the process of physical inheritance.

All organisms, including the very simplest, consist of two components, distinguished by Weismann as the "germ plasm" and the "soma." The germ plasm consists of the essential elements, or genes, passed on from one generation to the next; the soma consists of the body that may be produced as the organism. The distinction between the soma and the germ cells, propounded by Weismann in the germ plasm theory, emphasizes the role of the immortal, heredity-carrying genes and chromosomes, which are transmitted through successive generations. (excerpted from the *Encyclopedia Britannica*)

August Weismann's theory provides the basis for the general theory of the Life Cycle, which describes a bipartite system of germ cells and somatic body, the former containing the material of inheritance and the latter the body in which the germ cells dwell in part of the cycle. Weismann correctly assumed that the germ cells contain genetic chemicals, which at that time he could not identify, but which are now understood to be the genome. Nor did he recognize that the germ cells are in the form of a torus.

The Universal Torus

The premise of this book is that the form of the body originates in the toroidal structure of the germ plasm membrane, the observed shape of which accrues by amoeboid streaming, the process which causes the propulsion of the motile germ cells, resulting in their journey to the ovaries where they become eggs. This model attributes the formation of the body plan to the toroidal structure of the germ cell membrane and the supply of materials of construction to the genes contained within.

At every stage the multicellular body is in the form of a torus. The growing spherical nucleus causes the toroidal germ cell membrane to round up to become the egg cell which, in turn, is encompassed by a toroidal membrane system consisting of its three interconnected membranes: the outer plasma membrane, the intermediate endoplasmic reticulum, and the inner nuclear envelope. Cellular subdivision following fertilization creates hundreds of cells which form the blastula, a ball of cells, each of which is a part of the original egg cell membrane system. In the succeeding steps of gastrulation and organogenesis the motile cells, part of a morphogenetic field which originated in the germ plasm, reconstitute the toroidal form of the egg cell membrane and its predecessor, the germ cell, where the morphogenetic field originated.

The embryo is a torus, as is the succeeding adult body in all nature. To be even more specific, however, the underlying theme of all plant and animal form is observed to be the simple geometric figure called the multi-torus, a torus within a torus. It is demonstrated to be the result of the magnification of the observed multi-toroidal structure of the primordial germ plasm. While the Weismann germ plasm theory is a fundamental fact of biology, it is often overlooked in today's gene-centered scientific community.

Morphogenetic Fields

The Morphogenetic Field Theory is cited in this book as the basis for the apparent discrepancy between parts of the general model of morphogenesis presented and certain examples in observed nature, such as limb formation in vertebrates and insects. Morphogenetic fields are unseen blueprints which generate forms. The morphogenetic field is equivalent to an electromagnetic field that carries information, and is available through time and space without any loss of intensity after it has been created. Morphogenetic fields are created by the patterns of physical forms and help guide the formation of later similar systems where a newly-forming system is directed by a previous system by having within it a pattern that resonates with a similar pattern in the earlier form.

The morphogenetic field has been described and defined by many authors since the beginning of the century. Probably the most concise description of the morphogenetic field is discussed in a book authored by Scott F. Gilbert, *Developmental Biology* (2006), and excerpted here:

> ***The "Re-discovery" of Morphogenetic Fields***
>
> The 1920s and 1930s saw an enormous expansion of experimental embryology. The main objective of this science was the search for the rules of ordered form (*Gestaltungsgesetze*, in the German language in which most of the research was done). The concept that gave structure to this idea of ordered form was the **morphogenetic field.** This field was defined in various ways, but generally, it was that collection of cells by whose interactions a particular organ formed. The field had a definite boundary and the organ formed only from the interactions of cells within the field. Thus, these cells were specified in some way that told them they were members of the field. After the 1960s, the central paradigm was no longer the morphogenetic field, but the operon model of differential gene transcription. Even before then, in America, the morphogenetic field concept was in decline as embryology became based more in genetics than in physiology and anatomy. However, since the 1990s, the morphogenetic field concept has re-emerged, in part because molecular markers of these fields have been found.
>
> Needham (1950) approved of the use of fields to explain embryonic phenomena, and he combined the views of Spemann, Waddington, and Weiss in the following definition:

> *A morphogenetic field is a system of order such that the positions taken up by unstable entities in one part of the system bear a definite relation to the position taken up by other unstable entities in other parts of the system. The field effect is constituted by their several equilibrium positions. A field is bound to a particular substratum, from which a dynamic pattern arises. It is heteroaxial and heteropolar, has recognizably distinct districts, and can, like a magnetic field, maintain its pattern when its mass is either reduced or increased. It can fuse with a similar pattern entering with new material if the axial orientation is favorable. The morphogenetic gradient is a special limited case of the morphogenetic field.*
>
> What destroyed the morphogenetic field? One answer is that nothing destroyed the morphogenetic field. No data were presented arguing that the idea was wrong or that fields did not exist. Rather, the morphogenetic field was eclipsed and ignored.
>
> The morphogenetic field is returning as an important unit of development and evolution and phylogenesis. Moreover, it integrates the gene with evolution (Gilbert et al. 1996). Since genes work as part of pathways, and pathways are the physical interacting units of the morphogenetic fields, the field is the middle ground between the gene and evolution.

Scott Gilbert also coauthored with John M. Opitz and Rudolf A. Raff a landmark paper entitled, “Resynthesizing Evolutionary and Developmental Biology” (1996). The paper concludes:

> Just as the cell is seen to be the unit of structure and function in the body—not the genes that act through it—so the morphogenetic field can be seen as a major unit of ontogenetic and phylogenetic change. In declaring the morphogenetic field to be a major module of developmental and evolutionary change, we are, of course, setting it up as an alternative to the solely genetic model of evolution and development.

Certain phenomena exemplify the morphogenetic field, such as a flock of birds, a school of fish, a marching parade of soldiers, or a magnetic field. These are groups of elements which, when coming together, self-assemble in a recognizable configuration. After being disrupted by a barrier, the elements automatically reconstruct the original configuration. This phenomenon is demonstrated in a classical experiment in which sponge cells reassembled themselves after having been completely disassociated.

The necessary parameters for a morphogenetic field are a uniform substrate upon which individual elements in continuous motion are subject to an attractive and repellent force at the same time. That is, they are drawn together, but not too close together. In maintaining a moving formation for a period of time, elements learn their position in the mass, and after disruption can re-find their proper neighbors.

The morphogenetic field comprising the egg cell membrane, upon fertilization, is disrupted by cell division, by embryonic processes, and by gastrulation; the cells then reassemble during organogenesis, reconstructing the multi-toroidal form of the primordial germ plasm. Unaware of the underlying invisible organizing force of the morphogenetic field, the formation of organs seems like magic and has baffled embryologists to this day, inducing them to postulate the genetic theory and hold to it regardless of its acknowledged failure to produce a model after seventy-five years.

A Brief History of Morphogenetic Fields

The concept of the morphogenetic field, fundamental in the early twentieth century to the study of embryological development, was first introduced in 1910 by Alexander G. Gurwitsch. Experimental support was provided by Ross Granville Harrison's experiments transplanting fragments of a newt embryo into different locations.

Harrison was able to identify "fields" of cells producing organs such as limbs, tails, and gills, and to show that these fields could be fragmented or have undifferentiated cells added and a complete normal final structure would still result. It was thus considered that it was the "field" of cells, rather than individual cells, that were patterned for subsequent development of particular organs. The field concept was developed further by Harrison's friend, Hans Spemann, and then by Paul Weiss and others.

By the 1930s, however, the work of geneticists, especially Thomas Hunt Morgan, revealed the importance of chromosomes and genes for controlling development, and the rise of the new synthesis in evolutionary biology lessened the perceived importance of the field hypothesis. Morgan was a particularly harsh critic of morphogenetic fields since the gene and the field were perceived as competitors for recognition as the basic unit of ontogeny. With the discovery and mapping of master control genes, such as the homeobox genes, the pre-eminence of genes seemed assured. But in the late twentieth century the field concept was "rediscovered" as a useful part of developmental biology. It was found, for example, that different mutations could cause the same malformations, suggesting that the mutations were affecting a complex of structures as a unit, a unit that might correspond to the field of early twentieth-century embryology.*

* Wikipedia downloaded 3/30/09: "morphogentic fields."

Morphogenetic Fields Revisited

This book reveals a new approach to understanding the morphogenetic field theory, providing a graphic model of evolution and development. The discovery that the morphogenetic field is in the form of a multi-torus has led the present investigators to propose this new causative model of embryogenesis and evolution. Its universal predictability of living complex form places the model as a candidate solution to the problem of the origin of life.

The illustrations in this book show what at first glance are major discrepancies with observed nature, for example the well-studied steps of vertebrate and insect limb development, and the establishment of the insect gut tube. These exemplify the evolutionary phenomenon called "condensation" (Gould, *Ontogeny and Phylogeny*, 1977), where steps in ancestral sequences are absent in the embryology of descendants. Condensation is a major element in the phenomenon of recapitulation.

Thus, the sequences illustrated and labeled "achronological" are reconstructions of a theoretical morphology which accounts with accuracy for the gamut of possibilities in various limb configurations, which are presumed to have descended from actual ancestral events. The universal predictability of these sequences, well beyond coincidence, demands that they be considered plausible accounts of historical morphology, especially in the complete absence of any other reported model.

The Multi-Torus, Least Common Denominator of Complex Form

The finding that the embryonic tissue is in the form of a complex torus, or "multi-torus," was first published in 2003 by Jockusch and Dress. In 2007 Eric Davidson et al. published the discovery that genetic expression in the sea urchin embryo is in the form of a "dynamic torus." The configuration of one torus within another as the origin of form was first proposed by the author's laboratory. The multi-torus can originate as the outcome of the delamination of a toroidal bilayer, the form of the germ plasm. The study of biological form can be well illuminated by this remarkable topological phenomenon capable of simulating the forms which occur in animals and plants.

The dynamic toroidal surface is exemplified by the smoke-ring which is a ring torus composed of streaming laminations. The elongated tubular torus exhibits surprisingly different forms and behavior. The outer surface, under tension, is that of a regular tubular balloon. An internal canal passes from end to end, of star-shaped cross-section, the result of the compression of the tubular surface as it is forced to assume a minimal volume contained by the enclosing surface, like turning a sleeve inside out. The pattern can range from flat, like an empty fire-hose, to star patterns comprising three or more arms formed of two tightly clinging membranes surmounted by a tight circular tubule, which permits the containment and flow of fluid, separate from the interior volume. Radial animals and plants are of this cross-section.

The streaming toroidal membrane is subject to the continual change of lateral pressure from compression to tension as it passes from outside to inside. This process may be expected to engrain a longitudinal and circumferential pattern, subdividing the surface into quadrilateral segments. The membrane is robust enough to preserve the pattern under the compression of the internal passage, reconstituting it upon emerging at the other end. The deformation of this pattern establishes the phyletic bodyplan. Bending of the axis causes arc-shaped wrinkles in the outside ventral surface, morphological predecessors of limbs.

The delamination of a single toroidal surface creates the multi-torus of two or more concentric tori. The tense internal canal of the outer torus passes through the inner canal of the inner torus, constraining mutual rotation. Rotation results in the indentation of the outer membrane of the inner torus and can cause it to part, as in cell division.

These phenomena are the basis for the premise of the origin of complex form in the membrane of the germ plasm. The next pages describe other phenomena peripheral to the premise.

Phenomena Relative to Organic Self-Organization

Algorithmic and geometric modeling based on the torus can generate the full gamut of living forms. The question immediately arises: If these simple mechanically-driven steps are the process used by nature to create the living body, why was it not observed and reported by Aristotle, or by Leeuwenhoek in the seventeenth century? How come no mathematician ever discovered the algorithm by which the phlyetic forms can be generated by the topological deformation of a sphere?

As to the inability to observe the steps described in this model, the answer lies in the difficulty in the microscopic observation of the steps of early embryogenesis as well as the complexity of comparative embryology. Few conclusions can be drawn by the contemporary practice which focuses on merely five laboratory animal species. In fact, animal embryology occurs in a staggering variety of forms. Nearly a hundred different types of gastrulation are reported in nineteenth-century literature.

The focus of modern biology is biodiversity, the way species and individuals differ. The study is of the adult form of animals. Haeckel, von Baer, Goethe, and the nineteenth-century developmental biologists sought the origin of form in the way species are alike. Goethe sought a single Urform and Urpflanze from which all living form is derived. The answer was sought in the stages of embryological development.

A coherent train of mechanical events is interrupted by the intrusion of cleavage in the midst of the embryological sequence—the sudden subdivision of the egg into hundreds of cells. The theory that cellular subdivision is irrelevant to morphology clarifies the topological sequence that shapes the embryo. It is the basis of the once-prevalent organismal theory of development, which dismisses the role of the individual cells as mere bricks. This was replaced in the 1930s by the current developmental cell theory.

The Origin of the Phyletic Body Forms by the Deformation of the Multi-Torus

The worm, insect, crustacean and vertebrate body forms are determined by the topological exigencies inherent in the growth from a cylindrical, ellipsoidal, or spherical egg. The larval and adult bodies derive separately from the outer and inner toroidal membranes, respectively.

- *The elongated insect egg develops by the linear invagination of the ventral side caused by the lateral pressure of dorsal overgrowth.*
- *The curled, segmented body of the crustacean is delaminated from the surface of its round egg and embryo.*
- *The round vertebrate egg produces an embryo subject to an additional step consisting of the rotation of the inner torus within the stationary outer torus. The rotation causes the centric cable to press upon the dorsal surface, causing a deep groove which seals over, encapsulating the cable within. Ventral wrinkles form the limbs, the scapula detaches from the midline, forming the neck, and the pelvis from the lumbar vertebrae. The individual quadrilateral elements of the bilayer surface roll axially, forming bones, each encapsulated in a tube of muscle. Concurrently, the internal cables drag the outer torus behind it, turning it inside-out as its covers the inner body with the skin.*
- *The flower is simulated by the cylindrically constrained expansion of a multi-toroidal tube.*

(see Plate 11)

Toroidal Progenesis and Peristalsis

The chemical property of protoplasm to form a bilayer membrane leads to the formation of the streaming toroidal membrane configuration which is the progenitor of the basic biotic form of the segmented toroidal tube with paired appendages, as well as the physiological processes of axial locomotion by the coordinated movement of paired appendages and alimentation by the peristalsis of the internal gut tube.

Peristalsis is the periodic wave contraction of the muscles which comprise the tube of the gut, causing the movement of the contents of the gut from anterior to posterior. This phenomenon, universal in bilateral animals, may be accounted for as an evolutionary development of the streaming tubular, toroidal shape of the germ plasm membrane in ancestral worms. The membrane is composed of rows and columns of lipid molecules which organize themselves in segments in the shape of rings, from the front to the rear of the tubular animal body, exemplified in worms. The periodic waves of contraction of these rings cause peristalsis in the interior canal, and at the same time cause waves of successive deployment of the pairs of legs attached underneath the external tube.

Segmentation

The bilateral animals are almost universally segmented. The mechanical origin of segmentation has been studied for centuries. D'Arcy Thompson had summarized a number of plausible candidate causes already by 1917. In the subject model, the architecture of the shell or skull derives directly from the self-organized segmentation pattern of the hypothetical shell progenitor: a geodesically subdivided sphere. This is an elongated sphere segmented axially and longitudinally, including polar caps, the presumed natural pattern resulting from the compression failure of a sphere within a constraining sphere.

The spherical quadrilaterals of membrane defined by the segmental boundaries reduce in thickness in a gradient toward the center, which may be the locus of a hole which widens as the membrane is stretched. From the deformation of this pattern may be derived the shape and form of the shell and vertebrate skull. This occurs as the ventral hemisphere of the horizontally disposed figure is drawn into the dorsal hemisphere, producing a two-walled hemispheric dome. This structure is then drawn apart anterior-posteriorly. Except in turtles and horseshoe crabs, the posterior segments are lost, exposing a tubular or tapering tail. The pairs of holes at the anterior are the locus of the sense organs which occupy them by extrusion of the brain segments of the body membrane tube (see Plate 19).

Condensation

The comparison of the steps of early embryology of diverse species shows no consistent pattern. A universal event is gastrulation, which may be described as the partial entry of one part of the blastula into its own interior, at a point or along a line. There are a multitude of different ways this occurs in nature, all leading to the same result—the adult form. In observed embryogenesis, cells are seen moving about helter-skelter, but ending up in a seemingly predetermined arrangement, the embryo.

Insect eggs do not undergo cleavage yet display the morphological result of it. In the embryological phenomenon called condensation, the inheritance of embryology from an ancestor can include the loss of an early stage. The morphogenetic field has the property of trans-generational, morphological projection.

The embryo of the insect develops from a flat band on the ventral side. Cells from other places arrive to form a tube from the band, which will be the gut. There is no discernable guiding force which may be perceived in these seemingly random movements. Time-lapse photography shows the gut forming as both ends move to the center and join to form a tube, in concordance with the toroidal model.

The frequently cited model of condensation is the evolutionary predecessor of the modern winged insects, the wingless apterygotes which persist in silverfish species. Their embryo displays clearly that a tube is formed by the axial infolding of the band seemingly caused by a compressive force. This tube then disintegrates, the cells dispersing to other locations. Subsequently, these cells reorganize as the gut tube. In their descendants, the pterygote insects, the first step seems to be eliminated, leaving cells with an ancestral memory of their place in a tube.

The ability of cells to reassemble after dispersal is a well-known principle of cell behavior. This phenomenon can account for the ease with which cells seem to "know" their place in early embryonic development, a most baffling experience to behold. The cells are doing something they learned eons ago from ancestor species, inherited as a species-specific characteristic curvature of the membrane.

Regeneration

Humans are able to regenerate an eighth of an inch of lost fingertip, geckos an entire tail, some amphibians a leg. The lower animals can regenerate themselves in entirety from a small piece of tissue, the entire starfish from one arm. Regenerative power is lost with evolution toward complexity.

The fertilized egg cleaves. Both daughter cells each receive half of the egg cell membrane. Each eight-cell stage blastomere has an eighth of the membrane of the egg, and in many phyla, can generate the entire adult body. The subdivision of the egg and subsequent enlargement of the parts may be expected to form an enlargement of the structure of the egg in the form of the blastula. Blastulation is like dicing a beet. Nothing changes morphologically. Accidentally displaced blastomeres find their neighbors within hours.

Mechanical Induction

In a series of elegantly simple demonstrations, using advanced micro-manipulating and observational techniques, physicist Emmanuel Farge (2003) showed conclusively that simple distortive mechanical pressure on the embryonic cell is not only a sufficient condition, but also a necessary condition for gene expression. The theory of genetic control of cell division and evolution by random selection cannot be reconciled with the evidence of Farge's demonstration.

Farge induced gene expression in *Drosophila* egg cells by subjecting them to mechanical pressure. He then found that normal gene expression occurring in cells under pressure is blocked if the pressure is relieved artificially. In a third demonstration he restored pressure and observed that gene expression was restored as well.

The Farge discovery is the support for the theory that the embryo is formed by the undirected, step-by-step expansion of the tissue of cells. Cell division and gene expression are seen as the concurrent results of a third effect, the deformations in the membrane caused by cell crowding. Farge describes how β-catenin, torn loose from inside the membrane, heads for the nucleus where it instigates gene expression and cell division. Gene expression provides each daughter cell with growth factor to restore its full size. Increased crowding initiates another cycle in a self-propagating feedback loop. Cell crowding causes distortion, causes cell division, causes gene expression, causes cell growth, causes cell crowding.

Farge has demonstrated that mechanical force is a necessary as well as sufficient condition for gene expression in early *Drosophila* embryology. It follows that genes do not direct cell division. The fly's body form is not encoded in the genes. This refutes the widespread belief that the genome encodes shape. The genes make proteins which stimulate the growth of the cell pairs resulting from the division of the deformed cells that comprise naturally-occurring cell crowding patterns. The body makes the genes, not vice versa. The genome is a recording of historical developmental events, played back as a guide in each generation.

Cell crowding causes membrane deformation, which instigates the invasion of cytoplasmic β-catenin into the nucleus, initiating gene expression and cell division, resulting in further crowding and subsequent generations of cell division, in a self-propagating effect (Farge 2003). The premise that gene expression and cell division are both consequences of a third affect, that of mechanical induction, is the mechanism for this present model of development.

The Origin of Bilateral Symmetry

The higher forms of life are noted for symmetry, either radial or bilateral. The most outstanding feature of our own body is that it is composed of two mirror images, identical halves. That this is the result of identical growth rates of both sides is mechanically and physiologically improbable.

A simple explanation for this phenomenon is that the bilateral body originates ancestrally as a sphere with a horizontal polar axis, the dorsal and ventral sides bearing a gradient surface difference from dorsal to ventral, as though, for example, the result of exposure to overhead sunlight.

The sphere, by axial extension, results in a cylindrical tube of a like dorsal-ventral gradient. If the tube is parted at the ventral midline and the resulting surfaces separated, the result will be a bilateral figure of bilateral symmetry comparable to the human body.

If the midline parting is the result of a stretched elastic tissue, rather than a clean cut, then the extremities of the final surfaces will resemble our hands and feet, as though drawn apart, and the features of our faces, as though unities lying on the midline were split in two and drawn apart (see Plate 40).

Embryology vs. Genetics

Six hundred million years ago, dividing cells clung together in ever larger structures, following predictable steps which guide the folding and stacking of cells and tissue. A few million years later the Cambrian animals were completed. The following 450 million years were spent in changes in the proportions of the original phyletic body forms. The rise of the Cambrian forms has been a greatly debated phenomenon in biology and indeed so has the origin of form in general terms, with many biologists having attempted, and attempting still, to explain this phenomenon in a variety of ways.

Developmental Mechanics is the area of biology in which biologists, physicists, and mathematicians investigate the mechanical causation of biological phenomena, the school of thought epitomized by D'Arcy Thompson in his 1917 *On Growth and Form*. Swiss biologist Wilhelm His, considered the father of human embryology, wrote about mechanical causation in his great work, *Unsere Korperform und das physiologische Problem ihrer Entstehung* (1874), by replicating the form of the brain and other organs by the controlled axial compression of rubber and wax tubes.

The twentieth century was the century of genetics, culminating in the completion of the Human Genome Project in 2000. As the full sequence was unveiled, geneticist Walter Gilbert brashly asserted that scientists would now have access to "total knowledge of the human organism." The twentieth century passed as researchers failed to demonstrate that the genome contains a blueprint for body form.

The theory of mechanical causation grows out of work developed before the paradigm of genetics conquered biology. Today, geneticists emphasize the diversity of body plans, asserting that genetic differences give us the variety of shapes and forms in nature. However, it was Goethe who coined the term "morphology" as he searched for a single progenitor of form: "Tell Herder I have found the origin of flower form, and it may be the same for animals as well," (in a letter to his sister). The observation and study of form dominated biology in the nineteenth century; Cuvier and Geoffroy St. Hilaire debated whether the five "embranchments" were created separately or had common ancestry. In support of Goethe's single progenitor, Geoffroy famously described the insect as living inside its own skeleton. Knowledge beyond observation of the genesis of form came with the discovery of physical laws and the application of those to mechanics in the shaping of the embryo. There has always been a school of biologists who recognized that even organic matter should be subject to the physical and mathematical laws governing material in the universe.

Wilhelm His established the school of biology known as Developmental Mechanics (*Entwicklungsmechanik*) in the 1880s. He demonstrated the crucial role of a physical principle called "constrained expansion" in organogenesis. Constrained expansion occurs when a spherical or cylindrical surface such as that of an organ expands within an encapsulating inelastic envelope. His used controlled compression of rubber and wax tubes to replicate formation of the gut, brain, and other organs. Unfortunately, the work of His fell prey to the political atmosphere of the time.

Natural selection and its principle of "survival of the fittest" made Charles Darwin the darling of the Proto-Nazi eugenics movement, led by his powerful friend Ernst Haeckel. His' demonstrations of *Entwicklungsmechanik* refute natural selection (it was His who exposed Haeckel's infamous fraudulent embryo drawings). Haeckel prevented His' students from publishing or teaching in Germany. Wilhelm His' work on organic morphogenesis and the mechanics of development became history.

D'Arcy Thompson's *On Growth and Form* epitomized the study of mechanical causation of biological phenomena, but a new surge in genetics quickly overshadowed its principles. Choosing the fruit fly *Drosophila melanogaster* as a model organism, T.H. Morgan introduced developmental genetics in the early 1900s. Morgan made direct links from genes to mutations. His work quickly locked the scientific community into the paradigm of genetic determinism. Almost a century later geneticists relentlessly pursue the genetic blueprint for form in embryogenesis. The overarching school of thought ignores the fundamental role of mechanical forces in morphogenesis, thereby rejecting the rich, early research history of embryology.

On Stephen Jay Gould's *Ontogeny and Phylogeny*

In the previous pages the phenomenon of the morphogenetic field is recruited to explain how a pattern engrained in the rows and columns of the molecular membrane of the germ plasm is transferred to form the embryo. The first step transfers the pattern to the three contiguous membranes which enclose the cytoplasm of the egg—the plasma membrane, the endoplasmic reticulum, and the nuclear envelope—then, upon fertilization of the egg, to the resulting ball of a thousand individual cells comprising the blastula. Finally, these blastocysts assemble as if by magic, in the form of the embryo, guided by the "memory" in each of them of which its former and ultimate neighbor must be.

The observer of this embryonic ballet sees only the formation of the embryo, and cannot perceive the origins of the choreography that began eons ago. This is the wonderment of embryology. The great embryologists of the nineteenth century, von Baer and Haeckel, each accounted for this wondrously incomprehensible spectacle by presuming that there were missing stages which existed historically, and which are omitted for various causes. Haeckel's famous phrase "ontogeny recapitulates phylogeny," that is, embryology is a replay of evolution, sums up this immensely complex study which is dealt with in Gould's definitive work, *Ontogeny and Phylogeny* (1977).

Presented here are excerpts from Gould's book that should be sufficient for the purposes of this superficial treatment of ontogeny and phylogeny to account for the differences between the idealized drawings in the algorithm depicted in this book and that which the embryologist actually observes. At the same time, the extant political background of recapitulation theory may be perceived.

> I began this book as an indulgent, antiquarian exercise in personal interest. I hoped, at best, to retrieve from its current limbo the ancient subject of parallels between ontogeny and phylogeny. And a rescue it certainly deserves, for no discarded theme more clearly merits the old metaphor about throwing the baby out with the bath water. Haeckel's biogenetic law was so extreme, and its collapse so spectacular, that the entire subject became taboo; otherwise no

modern reviewer would begin with these words his account of a work that dared to mention it: "There are still those who would Haeckel biology" (Du Brul, 1971, p. 739).

But I soon decided that the subject needs no apology. Properly restructured, it stands as a central theme in evolutionary biology because it illuminates two issues of great contemporary importance: the evolution of ecological strategies and the biology of regulation. The starting point for a restructuring must be the recognition that Haeckel's theory requires a *change in the timing of developmental events* as the mechanism of recapitulation. For Haeckel, the change was all in one direction—a universal acceleration of development, pushing ancestral adult forms into the juvenile stages of descendants. Our current, enlarged concept does not favor speeding up over slowing down; all directions of change in timing are equally admissible. Paedomorphosis—the appearance of ancestral juvenile traits in adult descendants—should be as common as recapitulation.

Haeckel interpreted the gill slits of human embryos as features of ancestral *adult* fishes, pushed back into the early stages of human ontogeny by a universal acceleration of developmental rates in evolving lineages. Von Baer argued that human gill slits do not reflect a change in developmental timing. They are not adult stages of ancestors pushed back into the embryos of descendants; they merely represent a stage common to the early ontogeny of all vertebrates (embryonic fish also have gill slits, after all). (pp. 2–3)

◦

What … is the mechanism of recapitulation? *There is only one way to make recapitulation work under a theory of evolution by physical continuity. Every recapitulationist, from the staunchest Darwinian (Weismann) to the most militant neo-Lamarckist (Cope and Hyatt), upheld this mechanism; there is no other. It involves two assumptions.*

1. *Evolutionary change occurs by the successive addition of stages to the end of an unaltered, ancestral ontogeny.* This assumption provokes two problems. First, since many lineages involve thousands of steps, ontogenies will become impossibly long if each step is a simple addition to a previous ontogeny. Second, embryonic stages usually occur much earlier in time and

at much smaller sizes than the ancestral adult stage they represent. There must be some force continually operating to shorten ancestral ontogenies, thereby keeping the descendant's period of development within reasonable limits.

2. The length of an ancestral ontogeny must be continuously shortened during the subsequent evolution of its lineage.

I shall call the first assumption "the principle of terminal addition," and the second "the principle of condensation." (pp. 74–75)

Extra-Scientific Phenomena

Acupuncture Meridians and Bio-energy Fields

The panoply of extra-scientific forms of medicine has been a raspberry seed in the wisdom tooth of Western medicine. Thousands of MD's now practice acupuncture. Have billions of people for centuries been healed by a mere placebo effect? The difficulty rests with the insistence of modern medicine on a plausible theory for its existence. Modern genetic biology provides none. The theory presented in this book presumes to offer a scientifically plausible groundwork for one.

Although the primary purpose of this book is to report the phenomenon of the torus theory of morphogenetic fields as the origin of biological form, these same effects seem to bear relation to the whole series of phenomena recognized by the different branches of alternative medicine, generally rejected by science as lacking any substantive scientific causality. These have a common basis in what are called energy fields, patterned after the known phenomenon of electromagnetic fields which surround magnets and electrical conductors. Two categories of alternative medical modalities are discussed: meridian lines of acupuncture and bio-energy fields.

Meridian Lines of Acupuncture

The application of the torus theory of morphogenetic fields to bio-energy is apparent in the meridian lines which are the basis of acupuncture and acupressure. The traditional maps of acupuncture show meridian lines defining circuits which traverse the body, connecting points on the exterior surface with the interior locations of the internal organs, constituting a torus-shaped pattern.

The morphogenetic torus model claims that the pattern of the membrane of the germ plasm is the template of the body form. This membrane bears a pattern of meridian lines resulting from the streaming of this amoeboid cell as it traverses the body of the female of the species (ending up in her ovaries). While growing to form the egg (which, upon fertilization, will subdivide into cells and then quickly regroup to form the embryo), it will continue to maintain the fundamental meridianal alignment.

The regular rows and columns of the lipid molecules, of which the germ plasm cell membrane is known to be made, align themselves in a grid pattern dominated by anterior-posterior meridianal columns of molecules. These enter the interior at one end and exit at the other. They then flow around the outer surface again to re-enter in an eternal cycle. This observation provides a plausible, scientific origin for the otherwise inexplicable existence of the acupuncture meridians. The meridianal pattern of the morphogenetic field of the germ plasm may well be responsible for a resulting equivalent configuration of the cells of the adult body. [References at the end of this section provide access to other recent scientific attempts to account for acupuncture meridians.]

Bio-Energy Fields

Bio-energy is one of many known terms used by the adherents of alternative therapies to describe so-called energy fields that are thought to envelop every living being. It has been identified and acknowledged throughout the ages by different cultures in different parts of the world. Chinese acupuncturists call it chi, yogis in India use the terms chakra and prana. Wikipedia gives this brief description: "Nāḍi (the Sanskrit for "tube, pipe") are the channels through which, in traditional Indian medicine and spiritual science, the energies of the subtle body are said to flow. They connect at special points of intensity called chakras." Nadis seem to correspond to the meridians of traditional Chinese medicine. Other expressions commonly used to describe this type of energy include immanent energy, subtle energy, and magnetic energy. Many modalities of alternative medicine presuppose the existence of electric fields surrounding the body, and the presumption that these can be manipulated to generate well-being. That modern medicine rejects these techniques as unfounded has not prevented them from enjoying vast popularity, based on real results. The existence of electric fields within the body is an undeniable fact of science. The electric discharge accompanying cell division was observed and measured nearly one hundred years ago, reported by D'Arcy Thompson in 1917.

The nervous system is indeed an electric circuit. That any electric conducting circuit is surrounded by an induced electromagnetic field has been known since its discovery in 1830 by Michael Faraday. The morphogenetic field theory proposed in this book as the origin of form implies that each individual cell in the body has "knowledge" of its position within the tissues of the organism. The cell may recognize its neighbors by mutually compatible electromagnetic fields surrounding each cell. The coexistence of electromagnetic and morphogenetic fields provides a bona fide basis for the speculation as to the validity of manipulation by practitioners of various bio-energetic procedures, although real causes of described patterns have yet to be established. Thus, two areas of alternatives to modern medicine are provided with causative, scientific corroboration in the theory that form is derived from the meridianally patterned germ plasm membrane which becomes enlarged through cellular subdivision to guide the shape of the body. While this conclusion was not the original purpose of the lengthy investigations which led to the discovery of the origin of body form, it is certainly an interesting and useful by-product.

The Cell Electric

Much as the complex technological world is essentially electrical, so is the human body. The body's dependence on electricity and electromagnetic fields is far deeper than the simple electrical network which constitutes the nervous system. The elemental particle of the body, namely, the cell, consists of the cytoplasmic structure called the cytoskeleton, enclosed in a system of membranes, themselves a highly complex structure. The nature and functioning of the cell was examined in detail in the second half of the nineteenth century, enabled by the invention of the compound microscope and the microtome. This permitted the discovery and observation of the structure of the cytoskeleton and the action of the chromosomes during mitosis. The challenging quest for the understanding of this complex system was the preoccupation of many scientists in the new science of cell biology and experimental embryology.

The understanding of the cell taught in today's textbooks is the work of these many investigators whose names do not, for the most part, gain mention in their pages. Who hears of Rhumbler, Chambers, Bütschli, Hyde, Loeb, Whitman, Boveri, Fol, Strangeways, Hertwig, Lillie, Goodsir, Sedgwick, von Brücke, Nussbaum, Driesch, or Roux, to name but a few. This colossal body of investigation spanning over fifty years was summarized in 1917 in the 1,100 pages of D'Arcy Thompson's *On Growth and Form.* This venerated and neglected book is called the greatest work in science literature. Written at the beginning of the era which saw genetics come to dominate biological science, D'Arcy, as he is lovingly called by his many current admirers, saw the body as a holistic summum of the electrical energy field characteristic of the individual cell, as described by the vast body of scientists whose work he summarizes.

Thompson's concluding remarks in *On Growth and Form* on cell structure and the body as a whole (p. 345).

> Discussed almost wholly from the concrete, or morphological point of view, the question has for the most part been made to turn on whether actual protoplasmic continuity can be demonstrated between one cell and another, whether the organism be an actual reticulum, or syncytium. But from the dynamical point of view the question is much simpler. We then deal not with material continuity, not with little bridges of connecting protoplasm, but with a continuity of forces, a comprehensive field of force, which runs through and through the entire organism and is by no means restricted in its passage to a protoplasmic continuum. And such a continuous field of force, somehow shaping the whole organism, independently of the number, magnitude and form of the individual cells, which enter like a froth into its fabric, seems to me certainly and obviously to exist. As Whitman says, "the fact that physiological unity is not broken by cell-boundaries is confirmed in so many ways that it must be accepted as one of the fundamental truths of biology."

This distillate of his compendium of knowledge of the state of the art of cell science stands as a monument to the conception of the body as a dynamic energy field. It is the beginning of the understanding of the scientific underpinning of the otherwise poorly understood phenomena of non-Western medicine.

References

Berman, B., B. Pomeranz, and G. Stux. 1995. *Basics of Acupuncture.* Berlin: Springer-Verlag.

Dharmananda, S. 1996. *An Introduction to Acupuncture and How it Works.* Portland, OR: Institute for Traditional Medicine.

Kaptchuk, T. J. 1983. *The Web That Has No Weaver, Understanding Chinese Medicine.* Chicago: Congdon and Weed.

Lee, T-N. 2002. Thalamic neuron theory: meridians=DNA. The genetic and embryological basis of traditional Chinese medicine including acupuncture. *J. Med. Hyp.* 59(5): pp. 504–521.

Maciocia, G. 1989. *The Foundations of Chinese Medicine*, Edinburgh: Churchill Livingstone.

Matsumoto, K., S. Birch. 1988. *Hara Diagnosis: Reflections on the Sea.* Brookline, MA: Paradigm Publications.

Ross, J. 1995. *Acupuncture Point Combinations.* Edinburgh: Churchill Livingstone.

Shang, C. 2001. Electrophysiology of growth control and acupuncture. *Life Sci.* 68(12): 1333–1342.

Shudo, D. 1990. *Japanese Classical Acupuncture, Introduction to Meridian Therapy* (translation by Stephen Brown). Seattle: Eastland Press.

Stux, G., R. Hammerschlag, eds. 2001. *Clinical Acupuncture: Scientific Basis.* Berlin: Springer-Verlag.

Tsuei, J. J. Scientific evidence in support of acupuncture and meridian theory. Healthworld/Healthy.net. http://www.healthy.net/scr/Article.asp?Id=1087&xcntr=8 (accessed March 30, 2009).

The Periodic Table of the Elements

Hanging on the front wall of every chemistry lecture room and laboratory is a six-foot chart of the periodic table of the chemical elements discovered in 1869 by Dmitri Mendeleef. The periodic table is the monument of organization of the science of chemistry. The table accounts for the way chemistry works. It determines fundamental properties of the elements, for example, how many atoms of each element will combine with how many of every other element. The periodic table is a discovery, made by one man during an all night drunken stupor, a twice-told tale.

By the beginning of the nineteenth century fifty indivisible elements had been identified and the so-called law of triads showed that they occurred in groups of three with similar properties, i.e., chlorine, bromine, iodine; iron, nickel, cobalt; neon, krypton, argon; etc. By 1865 John Newlands reported to the Royal Academy that there was remarkable eight-fold periodicity in the list of elements arranged according to increasing atomic weight. One skeptical academician asked him wryly, "Mr. Newlands, have you tried to arrange the elements alphabetically?"

Mendeleef, working on the problem one night before a scheduled lecture to farmers on agricultural methods, played solitaire and drank vodka. In a drunken dream he saw the cards in thirteen numbers and four suits masquerading as the chemical elements. He awoke and made a deck of cards with an element on each. When he arranged them in order of atomic weight in rows of eight, he was amazed to see that the columns consisted of elements with like properties. He cancelled the lecture and wrote a paper for the academy that was accepted immediately.

The system he discovered was not universally acknowledged. What was it for? What did it prove? And there were two blank spaces. Ten years later, Gallium and Germanium were discovered with the exact properties predicted by the unfilled spaces. This was called proof of the efficacy of the table.

It is a fallacy to call this prediction the "proof." The table was a fact. The proof of the system discovered by Mendeleef lay in each of the entries and their coordinated position with all the others. The table constitutes successful predictivity beyond coincidence. If Gallium and Germanium had been discovered before he made the table, then would there be no proof?

The fact of Mendeleef's system demanded an account for it in the structure of the atom, then completely unknown. Twenty years later Rutherford, and then Niels Bohr, proposed a far-fetched hypothetical structure of the atom consisting of shells of increasing numbers of invented electrons which could account for the table. Although to this day no one has ever seen the structure, the atomic theory became the working law of chemistry and physics. If the complex of perfect order demonstrated by the table was a mere coincidence then the table would have meaning or consequences, much as if it was accomplished by arranging the element alphabetically. Human experience and judgment is the only test of the difference between meaningful order and mere coincidence. The discovery of an underlying hypothesis which can account for order is a necessary corroboration if not proof of meaningfulness.

Beyond Coincidence

The basis of the theory of form and evolution presented in this book begins with a discovery, followed by a hyphothesis to explain it. The illustrations show how a simple form simulates life by the application of compressive, or bending forces. The results may be questioned as being influenced by human intervention by an operator who knows what result is desired. The reader may judge whether the results are objective, or influenced by sleight of hand. The following provides a helpful rationalization of the otherwise overwhelming, dazzling panoply of species constituting biodiversity.

Biology has organized the millions of multicellular species as separate phyla, of which some thirty-six are recognized in the animal kingdom. Of these there are only two morphological kinds—the radial animals and the bilaterals. The radials, sea urchins, starfish and jellyfish types, are shaped like a sac. The bilaterals, including worms, insects, crustaceans, and vertebrates consist of one common structure, described as follows: a tube, segmented from front to rear, each segment bearing paired, transverse, jointed appendages attached from beneath, which act as limbs, mouthparts, and wings; a larger front-to-rear subdivision separating three regions, the head, thorax, and abdomen; a small tubular vessel running the length from front to rear containing nerve cells which communicate with the segments causing wave motion (peristalsis) which causes alimentation in the internal canal, and forward motion by coordinated movement of the limbs, and which terminates in the head with a pair of optically sensitive globes; and finally, a common method of sexual intercourse alike in earthworm, fly, bird, lobster, man, and woman.

Thus, there is only one kind of bilateral animal body form. Biodiversity is merely the myriad variants which occur on this single, immutable form. In the pre-"Silent Spring" days, when life teemed in abundance in fields, streams and city streets, this was apparent to kids and adults who could see that bilateral animals were a tube with legs.

The torus theory is demonstrated by dozens of plates of drawings which purport to account for virtually all phenomena of organic form: the shapes of the bilateral animal phyla, that is, worms, insects, crustaceans, and vertebrates, the radial animals, including jellyfish and starfish, the diverse, beautiful geometric forms of the microscopic diatoms, the wing patterns of butterflies, the stripe pattern shared by zebras and tigers, as well as the form of flowers. All these may be simulated by compressing or bending a torus.

If a plumber wants to bend a copper pipe he must do so slowly, and with care, otherwise it will suddenly crimp, forming an internal fold underneath the arc of the bend. Similarly, a tubular rubber balloon, when bent in an arc will suddenly form one, two, three, or more spaced internal, semi-circular, fan-shaped folds. The shape of the internal folds occur by mechanical forces instigated by the deformation of the tube and are beyond the control of the operator who does the bending. A tube made of a stretchable plastic material, when bent and crimped, and upon straightening, the fan-shaped internal fold will separate at the midline, and each of the halves created will rotate to a horizontal position, simulating the fins of a fish. These simple observations suggest the origin of paired limbs (see Plates 6–9).

Of the demonstrations presented, the most complex is the origin of the human skeleton, including the intricate form of the skull. Here, the maneuvers illustrated are based on the unique configuration of the multi-torus, one torus within another. The central canal of the outer torus, seen as a taut cable, passes within the internal canal of the inner torus. If the inner torus rotates with respect to the outer torus, the taut, central cable will deeply indent the surface of the inner torus from the upper lip upwards, burying itself in the skull, and beneath the surface of the back, forming the configuration of the nerve chord, as the segmented back regions closes over it, forming the vertebrae. The illustrations show that one and a half rotations of the inner torus can simulate the vertebrate skeleton form, including the common stripe pattern of tigers and zebras (see Plates 20 and 23).

The same torus, which can create members of the animal kingdom, when grasped by a tightening collar at one end, causes a bulge below that simulates the radially segmented ovary of the flower, the configuration of the succeeding fruit. Upon the release of the compressed upper part, it creates in perfect detail the anther, pistils, and petals of the flower. This is confirmed when you cut an orange in half, and observe the radial segments and the central star-shaped pith, in accordance with the prediction of the algorthmic model (see Plates 43 and 44).

This presents the case that these effects are beyond the control of the operator who applies simple forces, without constituting directed human intervention in their outcome. The universal predictivity of the torus model may well be claimed to be beyond coincidence, and suggestive of an underlying cause. The theory presented is that the underlying cause is in the toroidal form encoded in the membrane of the primordial germ plasm. That there is a plausible cause obviates the premise that the whole thing is a mere coincidence.

II

The Self-Organization of Biological Form

The central premise of this book is expressed in the following pages in the form of a scientific paper.

The Self-Organization of Biological Form

The Simulation of Phyletic Form by the Deformation of a Toroidal Surface

ABSTRACT

This paper reports the discovery of an organic form simulator consisting of a toroidal surface that, on mechanical deformation, assumes the archetypal forms of various plant and animal phyla, with substantial predictive success. Mechanical models of latex and synthetic polymer film were fabricated which, upon inflation, demonstrate topological congruency of the forms that they assume with those of plant and animal phyla. Since the precursor of the egg, the primordial germ plasm, is well known to be a streaming toroidal membrane, the paper examines the possibility that natural organic form is derived from the self-organized structure of its membrane.

KEYWORDS

torus; primordial germ plasm; pseudopodal locomotion; sol-gel; morphogenetic fields

INTRODUCTION

The patterns seen in the organization of living nature have engendered wonderment since Aristotle as to the origin of the geometrically regular shapes and forms seen in shells and in the surfaces of many species. Attempts at the mathematical rationalization of form flourished in the idealizing age of eighteenth-century reason, following von Haller, Bonnet, Oken, and epitomized by Johann Wolfgang von Goethe, who saw the living form as a dynamic system built on a single formative principle. In *The Metamorphosis of Plants*, Goethe postulates that all flower parts are derived from leaves. He coined the term "morphology." The mathematical rationalization of natural form is exemplified by the form-encyclopedist D'Arcy Thompson, and more recently by Stephen Wolfram (Thompson 1942; Wolfram 2002).

Goethe famously proclaimed his discovery of the Urpflanze as the Protean progenitor of all botanical form. He failed to produce a model and his theory of plant metamorphosis has not survived the test, but his approach to natural science fathered the search for the ideal form that directed goals of the nineteenth-century developmental biologists. These morphologists rationalized the vertebrate skeleton, including the accepted evolutionary phylogenetic accounts such as the visceral arches in fishes and the complex head, ear, and neck bones of later vertebrates, which are still part of classical anatomy and developmental biology. For most of the last century the search for the origination of organismal form has come to be sought at the molecular level.

MATERIALS AND METHODS

Recent conclusions as to the toroidal nature of the embryonic tissue are published by Beloussov and Grabovsky 2006; Kraus 2006; Presnov, Malyghin, and Isaeva 1988; Davidson 2008. Jockusch and Dress (2003) determined that the embryonic membrane is a multi-torus. This led to this investigation of the morphological properties of the amoeboid pseudopod, for which purpose toroidal balloons of latex rubber, polyvinylchloride, and other elastomeric sheet materials were fabricated in varying sizes and proportions (Plate 1). Small (10 cm) toroidal balloons made as water toys, widely commercially available, were used. These were filled with air, water, glycerine, honey, and other fluids of varying viscosity. A 20 cm balloon, filled with water, and then submerged in water, was observed to assume a series of distorted configurations which bear resemblance to various developmental forms of animal and plant species by simple manual manipulation of the hydrostatic pressure in the parts of the structure. Schematic drawings were made of predictable extrapolations of the deformations. These drawings are observed to corroborate the congruence of the model with the adult forms of the species suggested in the experiment.

Observations of artificial pseudopod models demonstrate that slow toroidal streaming of a bendable, elastic membrane causes the formation of axial folds with tubular edges in the surfaces of the interior canal, which subdivide the figure radially. These are caused by lateral forces as the cylindrical surface enters the reduced space occupied by the interior surface of the torus. The internal canal of the axially extended torus may assume a radial star pattern of different numerical symmetries, depending on simple physical parameters, such as the thickness and elasticity of the sheet. The models used have four-fold symmetry, creating in cross section a double-walled cross with a circular extremity. The two inflated flexible co-axial tori assume a single stable position with respect to each other. The inner cruciform canal of the outer torus passes through the center of the cruciform inner canal of the inner torus. Mechanical forces rotate the tori to fit the spaces, bringing their axes 45 degrees apart. The cross section of the multi-torus shows two crosses, one rotated 45 degrees with respect to the other, in eight-fold symmetry.

EXPERIMENTAL DEMONSTRATIONS

Constrained Expansion

The expansion of a spherical surface within a fixed sphere produces predictable, internally directed growth of indentations, wrinkles, and folds. The topological phenomenon giving rise to the embryonic shape is that of the constrained expansion of an inflated, flexible, enclosed surface. In the growth scheme of nature new growth of internal layers of the body enclosure cause the outer layers to molt, as in insects, or to shed gradually in bathing, as in humans. The following demonstrations show that the phyletic forms are the consequence of the expansion of the multi-torus (Plate 3).

Axial Segmentation

The model elastomeric tube, water-filled and submerged, is observed to form uniform sinusoidal circumferential corrugations for most of its length. This phenomenon is described by D'Arcy Thompson (1917) in his study of the causes of segmentation in the animal body form.

The Self-Organization of the Toroidal Surface

The form assumed by a membrane-bound colloidal fluid is exemplified in the phenomenon of organic amoeboid locomotion, well known to be the result of toroidal streaming and understood as a sol-gel inter-phase phenomenon. The transparency of moving fluid surfaces makes observation indistinct. The internally directed expansion of a spherical surface within a fixed sphere produces a curled, axially segmented tube with internally directed creases on the concave side, repeating the configuration of the larval stage of the bilateral phyla (Plate 55).

Amoeboid locomotion is simulated by the gravity descent on an inclined plane by a model toroidal balloon (Plate 1, Fig. 1). This paragraph presents an idealized model of pseudopodal locomotion based on the classic descriptions of the phenomenon: The protoplasm mass generates an enveloping membrane. A weakness in the wall results in a circular port from which newly-formed membrane radiates as it emerges, causing a domed cylindrical tube to project from the surface. Sol state material exits the port immediately to undergo gelation, producing more membrane which radiates to flow backwards over the body, re-entering at the posterior port.

While it has not been clearly observed, it is theorized that the membrane as it is compressed laterally on entering the interior space will wrinkle in a radial pattern like the one formed when a coat sleeve is turned inside out. In the mechanical models, the pattern produces radial segmentation and forms pinched-off tubes from the distal borders of the axial segment walls that envelop the tubular figure from posterior to anterior. These act as sealed tubes and are capable of conducting the flow of water axially, separately from the fluid content of the cell. Squeezing the middle of the model can pump water back and forth through these peripheral tubes, in simulation of the action of the heart.

Balloon Topology

The tubular toroidal balloon is observed to be subject to three types of failure under axial tension and bending torque. The first of these is an observed pattern of circumferential undulations that segment the surface axially (Plate 1, Fig. 3). The second is failure by small, infolded bands on the concave curve of the bend (Plate 1, Fig. 4). The third failure is the subdivision of the tube in a small number of large folds, the familiar bending maneuver of a tubular balloon (Plate 1, Fig. 5). A bending torque on the axis produces a series of quantified stages of folded forms, rather than a continuously variable bend. This phenomenon is familiar to anyone who has manipulated a tubular balloon where axial torque produces a succession of separated stages of equilibrium rather than a smooth bend.

This is the common reaction to the bending of an ordinary tubular party balloon. The interior folds of the bend are observed (Plate 12, Insect Fig. 3) to be in the form of lunettes, in the shape of the sector of a circle, comprising two tightly adhering surfaces bordered at the top rim by a thin, straight horizontal tube, the predictable folding pattern of an elastic sheet.

It is noted that the axial deformation of a tubular balloon results in the folding rather than in the bending of the tube. The folds occur in quantified instantaneous sequences with no intermediate forms perceptible. This quantification of toroidal form implies a limited number of morphological niches, with no in-between stages.

Limbs

The model provides for the following unobserved, hypothetical steps of early embryogenesis, presumed to have been lost ancestrally.

Bilateral animals may generate two, three, four, or five pairs of creases, or internal folds, in the ventral hemisphere (Plate 9). These result in paired limbs and mouthparts. In insects, two large lunettes occur which extend into the dorsal hemisphere, forming wings (Plate 15a). In lepidopterans, these are so large that they extend, super-imposed over the paired limb primordia and impress a pattern on the wing primordia which results in the wing spots. The compressive force of the pairs of internal axial tube structures on either side impresses the side bar pattern. The scalloped edge arises as the two halves of the wing primordia are bisected and stretched apart at the midline (Plates 15a, 15b). Anterior-posterior tension in the outer tube produces circumferential undulation along the length of the figure (Plate 15a). These mimic axial segmentation, universal in animal form (Thompson 1942). The streaming of an inflated toroidal surface produces a unique terminal structure, counterintuitive in shape and behavior, and difficult to conceive without the experience of personally handling a model artificial pseudopod.

DISCUSSION

The individual bilateral animal is toroidal in all stages, from germ plasm to egg cell membrane, embryo, larva, and adult. The pattern established in the germ plasm membrane is the precursor of the egg cell membrane, which conserves the patterned structure in the plasma membrane and the endoplasmic reticulum. In phyla that experience subdivision of the egg by cleavage, the resulting blastula and embryonic membrane is a mosaic enlargement of the germ plasm structure. Cellular subdivision disrupts the morphogenetic field inherent in the germ plasm tissue. In organogenesis the cells are seen moving to reassemble the pattern by finding their neighbors. Cellular subdivision, by this standard, is morphologically irrelevant. It has served to conceal the underlying formative imperative in the plasm tissue, by shuffling the deck. The activity of the genes in providing materials of construction at timed intervals maintains, and occasionally changes, the proportions of the otherwise immutable phyletic body-plan over generations, but does not encode for the form itself (Gould 1977).

Organogenesis

Given the self-organized multi-torus structure, the body form is established in two sudden, consecutive, topological events. First, the establishment of the germ layers by the expansion of the toroidal membrane, next the "gastrulation," or stomach shape formation of the membrane resulting from the incursion of the entire fluid content of the lower half of the body into the upper half.

Self-Organization in Oogenesis

The egg of many insect orders is in the form of a longitudinally segmented ellipsoid, the segments separated by internally directed membranes which at a later stage may extend partially or fully toward the center. In insects cellular subdivision occurs well after the body form takes shape in the egg membrane. The delamination of the bilayer membrane of the torus can result in the formation of a second torus within the first. These may move freely with respect to each other, constrained only by the internal canal of the outer torus that passes within the internal canal of the inner torus.

The Primordial Germ Plasm

The discovery of the germ plasm, the amoeboid predecessor of the egg, occurred at the end of the nineteenth century by investigators who sought the mechanism of inheritance, leading to August Weismann's famous theory of 1878 that inheritance results from the so-called immortal germ line cells which dwell in the tissue of mortal, somatic bodies.

The immortal germ plasm divides periodically, one sister-half dwelling within the host somatic body resulting from the growth of the other sister, exiting before she dies, to repeat what is known as the Life Cycle.

Life Without Evolution, Form Without Genes

Conjoined twins and other anomalies have unique structures including novel bone forms, muscles, nerves, and blood vessels all organized into a perfect working system to accommodate the new form. This complex of organs, each fitted perfectly for its function, appears all at once, perhaps for the first time in the history of the species. This occurs without the benefit of evolution or any special code in the DNA.

Hence, complex form does not require a genetic code. Substantial modifications of form can occur in a single generation.

CONCLUSIONS

The paper demonstrates that the forms of nature can be simulated by the deformation of the topological figure called the multi-torus. Since the living germ plasm is in the form of a toroidal bilayer, that nature makes use of the algorithm in the creation of form is a plausible solution to the problem of morphogenesis.

References

Arber, A. 2002. *Goethe's Botany.* United States: Sacred Science Library.

Ball, P. 1999. *The Self-Made Tapestry: Pattern Formation in Nature*. Oxford: Oxford University Press.

Bard, J.B.L. 1977. A unity underlying the different zebra striping patterns. *J. Zool.* (London) 183: 527–539.

______. 1981. A model for generating aspects of zebra and other mammalian coat patterns. *J. Theor. Biol.* 93: 363–385.

Beloussov, L., and V. Grabovsky. 2006. Morphomechanics: Goals, basic experiments and models. *Int. J. Dev. Biol.* 50: 81–92.

Ellers, O., and M. Telford. 1992. Causes and consequences of fluctuating coelomic pressure in sea urchins. *Biol. Bull.* 182: 424–434.

Farge, E. 2003. Mechanical induction of Twist in the *Drosophila* foregut/stomodeal primordium. *Current Biology* 13: 1365–1377.

Gould, S.J. 1977. *Ontogeny and Phylogeny*. Cambridge, MA: Belknap Press.

Jockusch, H., and A. Dress. 2003. From sphere to torus: A topological view of the metazoan body plan. *Bulletin of Mathematical Biology* 65: 57–65.

______. 2004. Letter to the Editor. *Bulletin of Mathematical Biology* 66: 1455.

Kamm, K. et al. 2006. Axial patterning and diversification in the Cnidaria predate the Hox system. *Current Biology* 16: 1–7.

Keller, R., L. Davidson, and D. Shook. 2003. How we are shaped: The biomechanics of gastrulation. *Differentiation* 71: 171–205.

Kraus, Y.A. 2006. Morphomechanical programming of morphogenesis in Cnidarian embryos. *Int. J. Dev. Biol.* 50: 267–275.

Maresin, V.M., and E.V. Presnov. 1985. Topological approach to embryogenesis. *J. Theor. Biol.* 114: 387–398.

Murray, J.D. 1981. A pre-pattern formation mechanism for animal coat markings. *J. Theor. Biol.* 88: 161–199.

______. 1989. *Mathematical Biology*. New York: Springer.

Nishii, I., S. Ogihara, and D. L. Kirk. 2003. A kinesin, InvA, plays an essential role in *Volvox* morphogenesis. *Cell* 113: 743–753.

Oparin, A.I. 1952. *The Origin of Life*. New York: Dover.

Painter, K.J. 2000. Modeling of pigment patterns in fish. In *Mathematical Models for Biological Pattern Formation* (Eds. P.K. Maini and H.G. Othmer), IMA Vols. in Mathematics and its Applications 121, 59–82. Berlin: Springer.

Pauchard, L., and Y. Couder. 2004. Invagination during the collapse of an inhomogeneous spheroidal shell. *Europhysics Letters* 66(5): 667–673.

Perales-Graván, C., and R. Lahoz-Beltra. 2004. Evolving morphogenetic fields in the zebra skin pattern based on Turing's morphogen hypothesis. *Int. J. Appl. Math. Comput. Sci.* 14(3): 351–361.

Presnov, E.V., S.N. Malyghin, and V.V. Isaeva. 1988. Topological and thermodynamic structures of morphogenesis. In *Thermodynamics and Pattern Formation in Biology* (Eds. I. Lamprecht and A.I. Zotin), 337–370. Berlin: Walter de Gruyter.

Presnov, E.V., V.V. Isaeva, and A. Chernyshev. 2006. Topological patterns in metazoan evolution and development. *Bulletin of Mathematical Biology* 68(8): 2053–2067.

Seilacher, A. 1979. Constructional morphology of sand dollars. *Paleobiology* 5(3): 191–221.

______. 1985. Discussion of Precambrian metazoans. *Phil. Trans. Roy. Soc. London B.* 311: 47–48.

______. 1989. Vendozoa: organismic construction in the Proterozoic biosphere. *Lethaia* 22: 229–239.

______. 1992. Vendobionta and psammocorallia: Lost constructions of Precambrian evolution. *J. Geol. Soc.* 149: 607–613.

______. 1997. *Fossil Art.* Drumheller, Alberta: The Royal Tyrrell Museum of Paleontology.

______. 2007. *Trace Fossil Analysis.* Berlin: Springer.

Seilacher, A., and R. Hauff. 2004. Constructional morphology of pelagic crinoids. *Palaios* 19(1): 3–16.

Schmitt, R., and M. Sumper. 2003. How to turn inside out. *Nature* 424: 499–500.

Sheldrake, R. 1981. *A New Science of Life: The hypothesis of formative causation.* Los Angeles: J. P. Tarcher.

Shoji, H., and Y. Iwasa. 2003. Pattern selection and the direction of stripes in two- dimensional Turing systems for skin pattern formation of fishes. *Forma* 18: 3–18.

Smith, J., C. Theodoris, and E.H. Davidson. 2007. A gene regulatory network subcircuit drives a dynamic pattern of gene expression. *Science* 318(5851): 794–797.

Thompson, D.W. 1992. *On Growth and Form*, the complete revised edition. New York: Dover.

Turing, A. 1952. The chemical basis of morphogenesis. *Philos. Trans. Roy. Soc. London B.* 237: 37–52.

Wolfram, S. 1984. Cellular automata as models of complexity. *Nature* 311: 419–424.

______. 1984. Universality and complexity in cellular automata. *Physica D.* 10: 1–35.

______. 2002. *A New Kind of Science*. Champaign: Wolfram Media.

III

Plates, Related Papers, and Publications

Plates

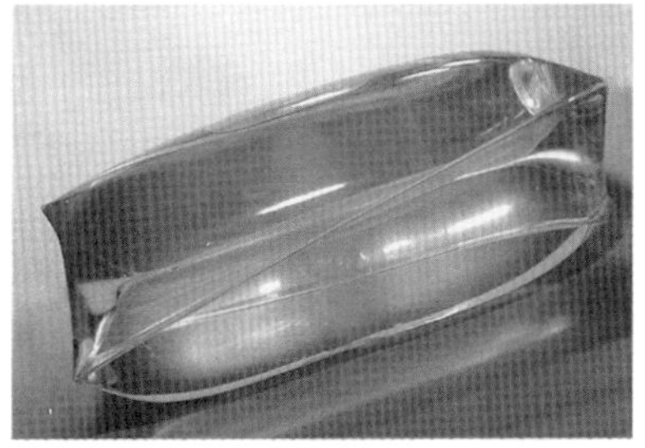

FIG. 1

FIG. 2

FIG. 3

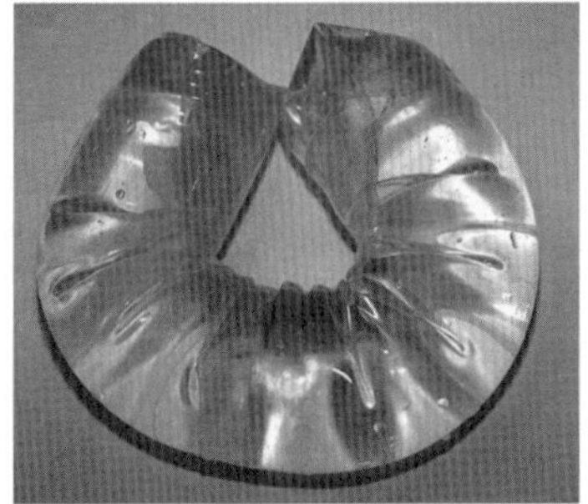

FIG. 4

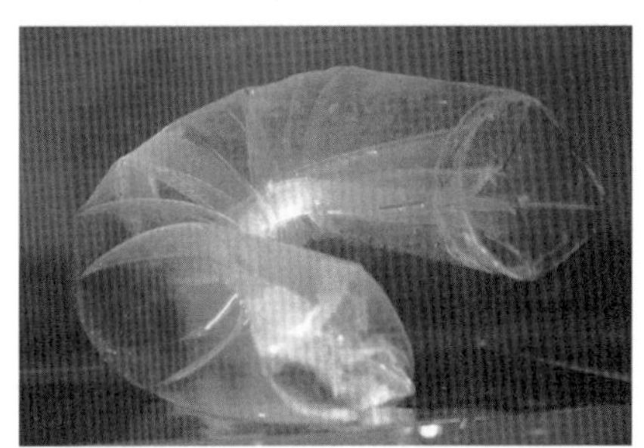

FIG. 5

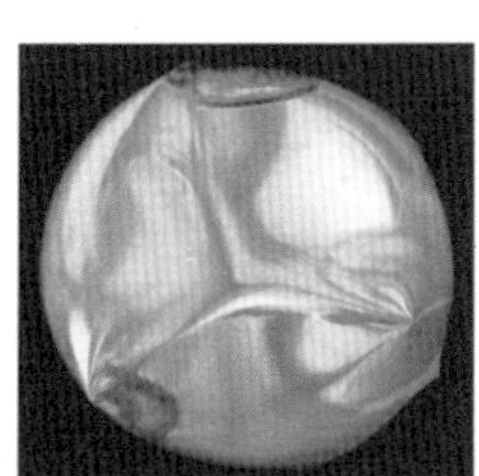

FIG. 6

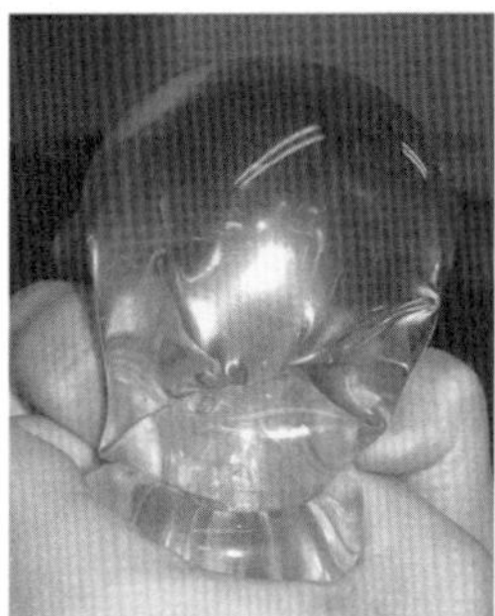

FIG. 7

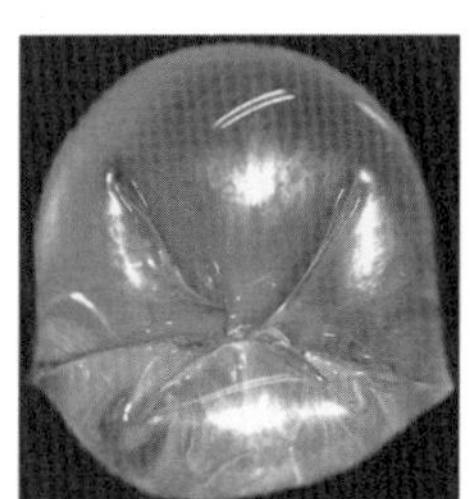

FIG. 8

WATER-FILLED TOROIDAL BALLOONS DEMONSTRATING MORPHOGENESIS

PLATE 1

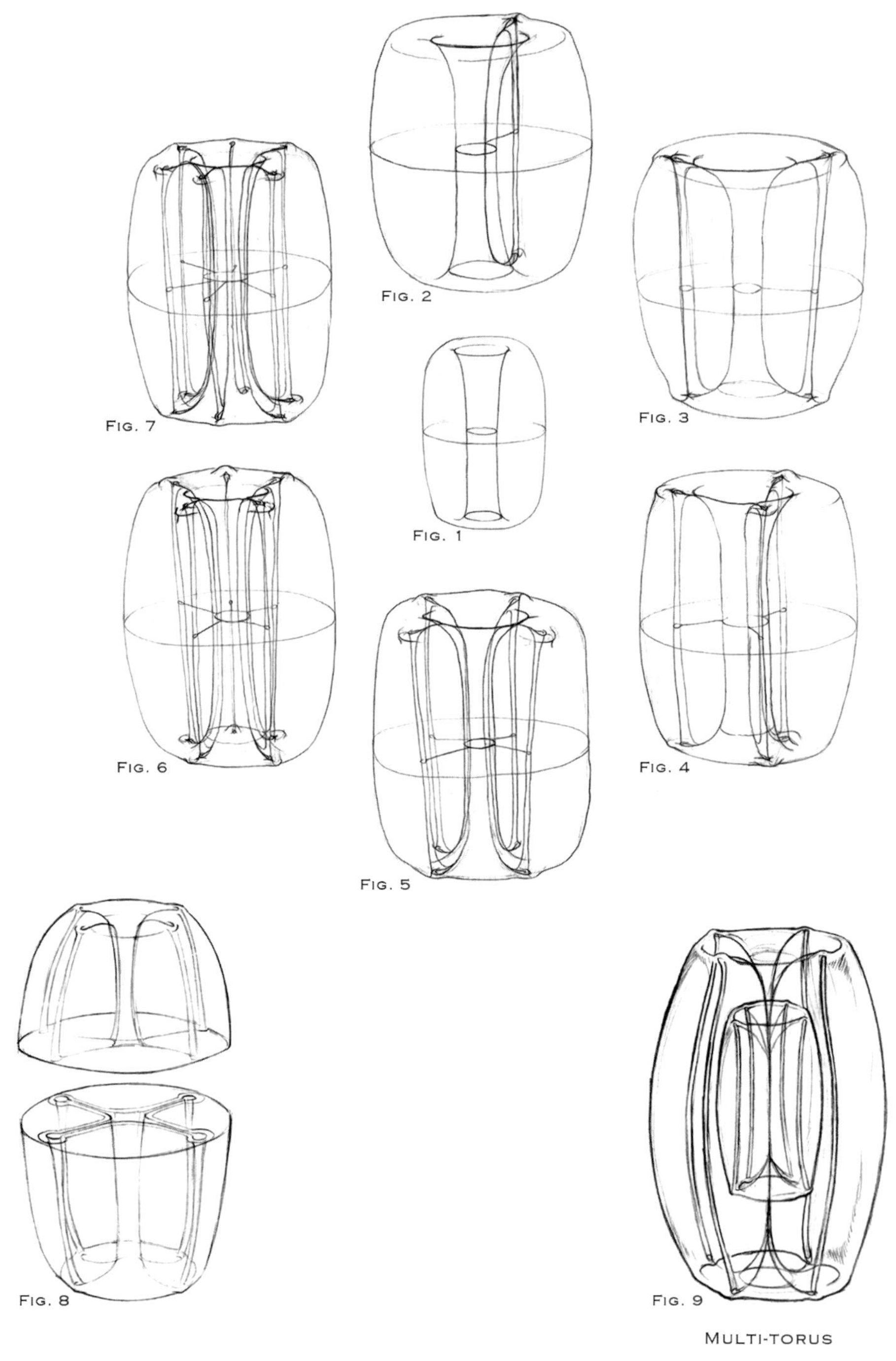

TOPOLOGICAL DIVERSITY OF THE TOROIDAL SURFACE

PLATE 2

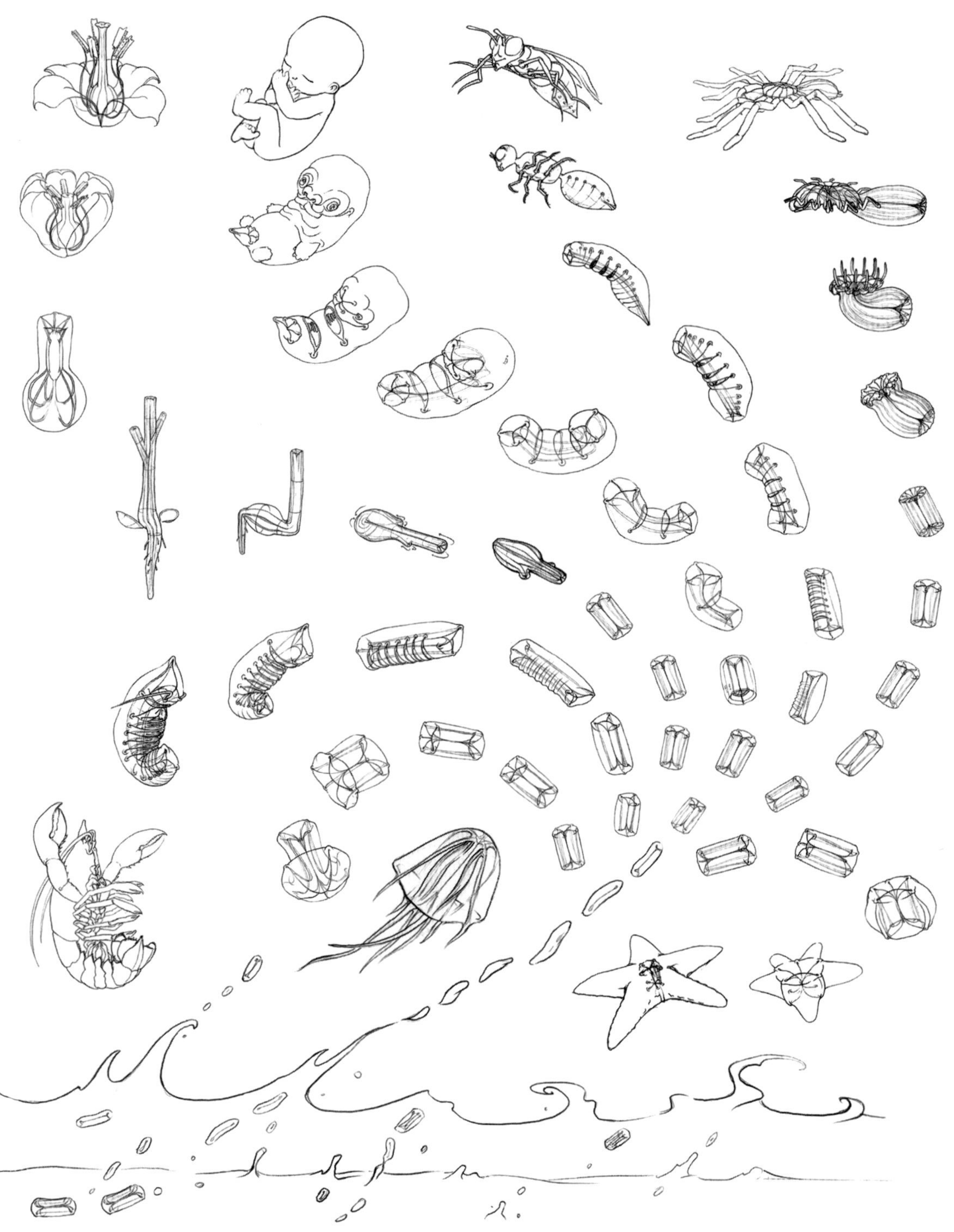

Flowchart of Phyletic Differentiation by Mechanical Deformation

Plate 3

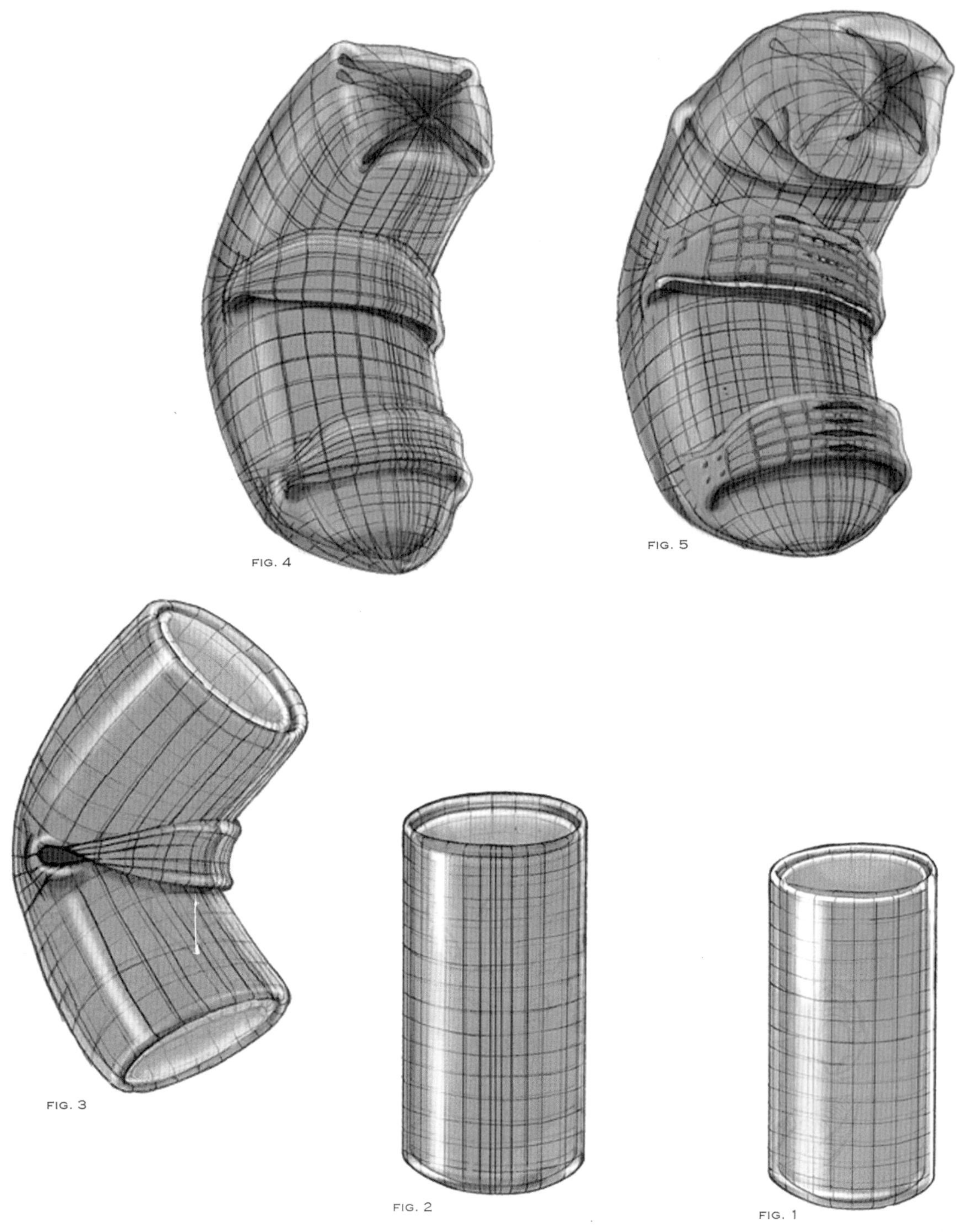

Vertebrate Limb Development (Achronological)

Plate 4a

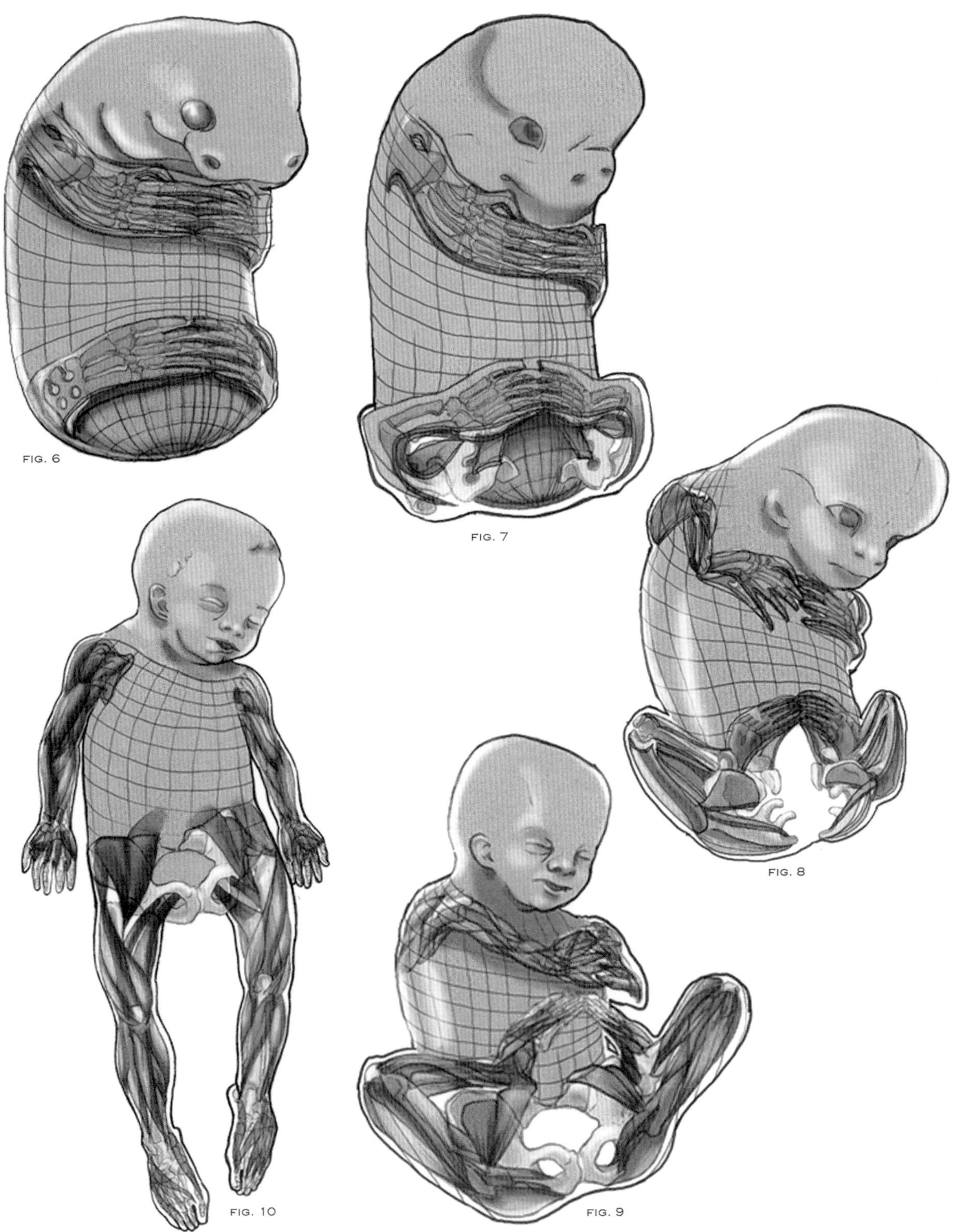

Plate 4b

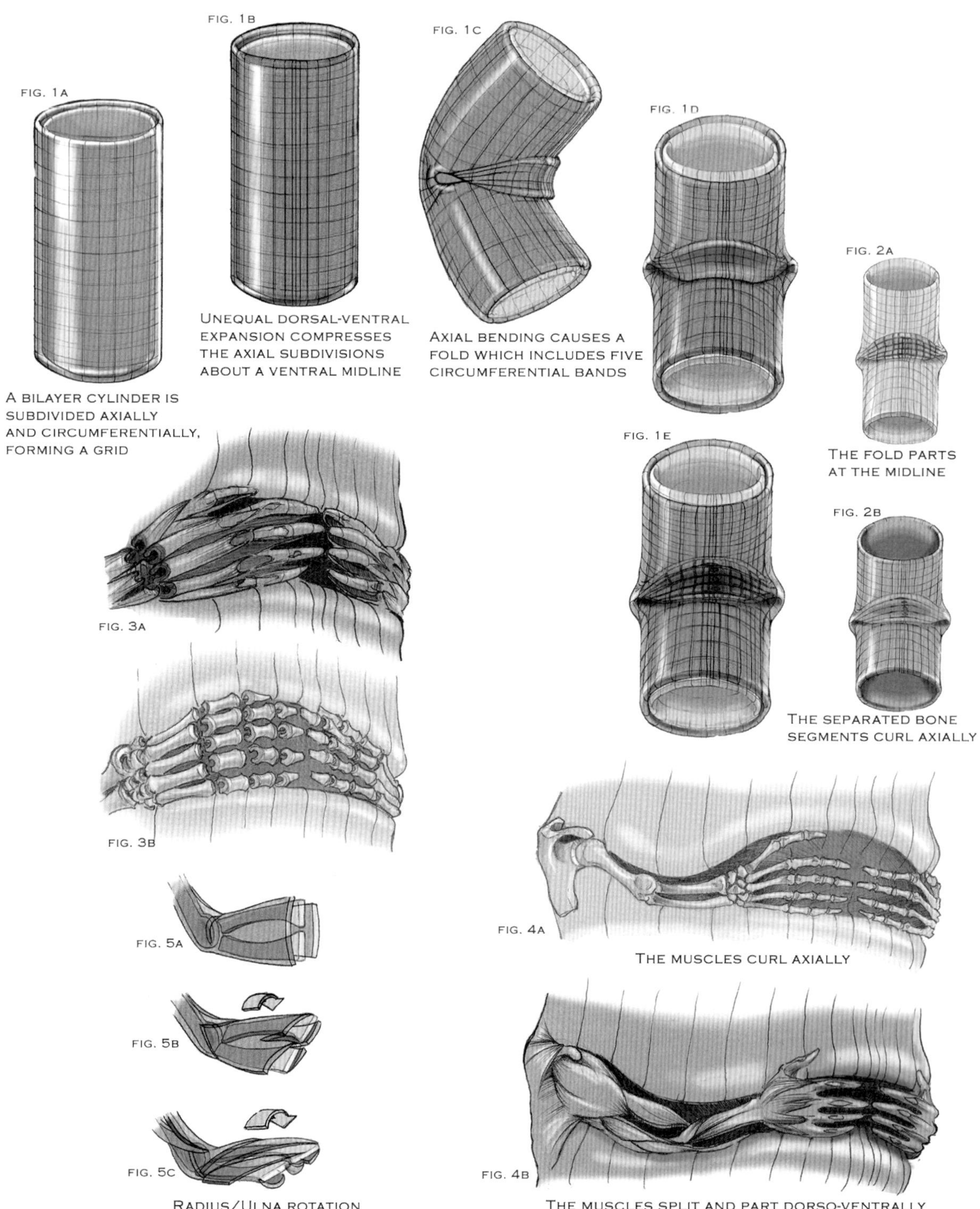

VERTEBRATE LIMB DEVELOPMENT (ACHRONOLOGICAL)

PLATE 5

FIG.3 BELOW SHOWS THE OBSERVED STAGES OF VERTEBRATE LIMB FORMATION TO BE COMPARED WITH THE THEORETICAL MORPHOLOGY PRESENTED IN THE PLATES OF STAGES WHICH ACCOUNT GEOMETRICALLY FOR THE ADULT LIMB. THESE STAGES PRESUMABLY OCCURRED IN THE LIMB PRIMORDIA OF THE VERTEBRATE ANCESTOR, AND HAVE BEEN LOST BY EVOLUTIONARY CONDENSATION (GOULD 1977), BUT ARE RETAINED AS MORPHOGENETIC FIELDS.

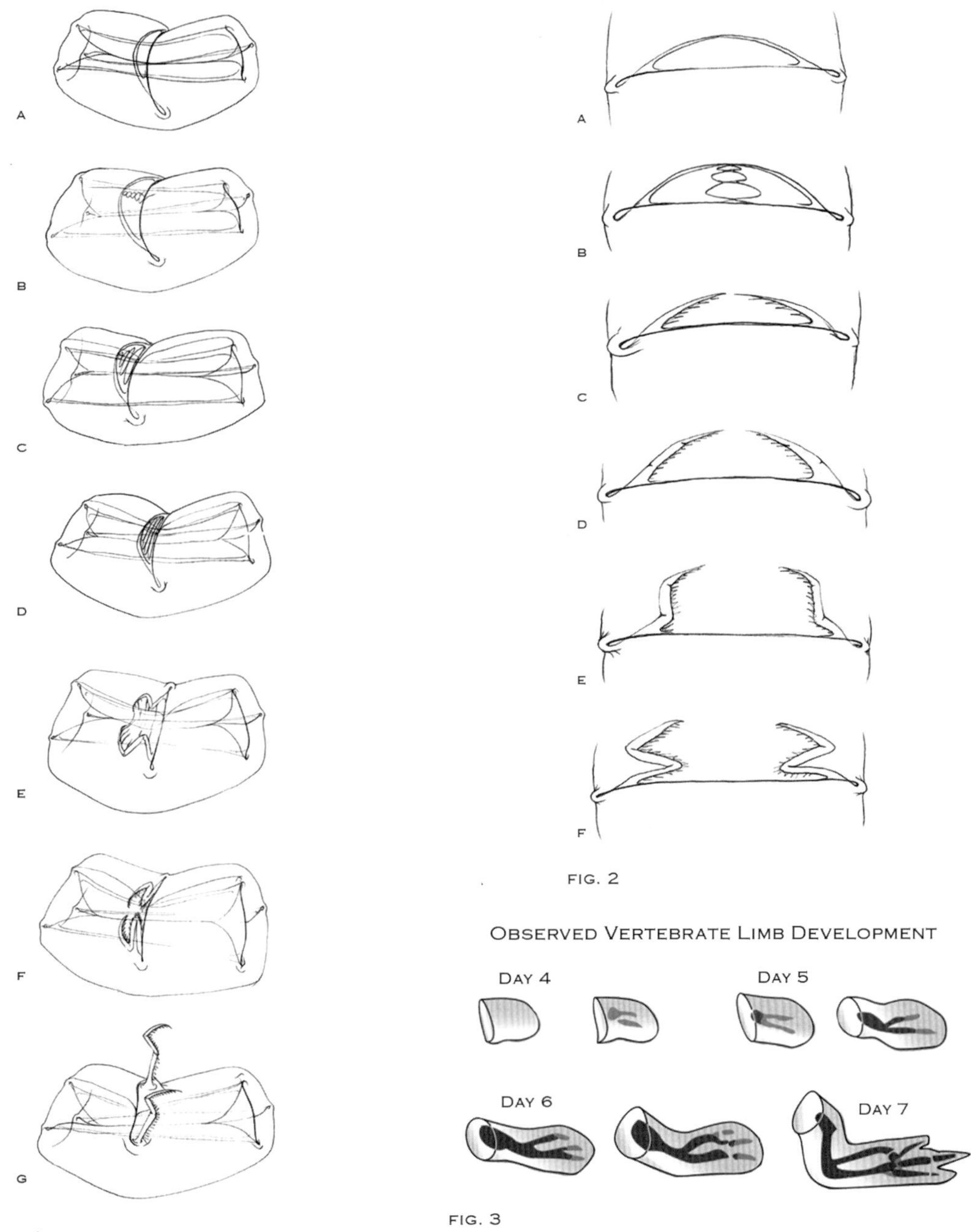

VERTEBRATE AND INSECT LIMB DEVELOPMENT BY MORPHOGENETIC FIELDS

PLATE 6

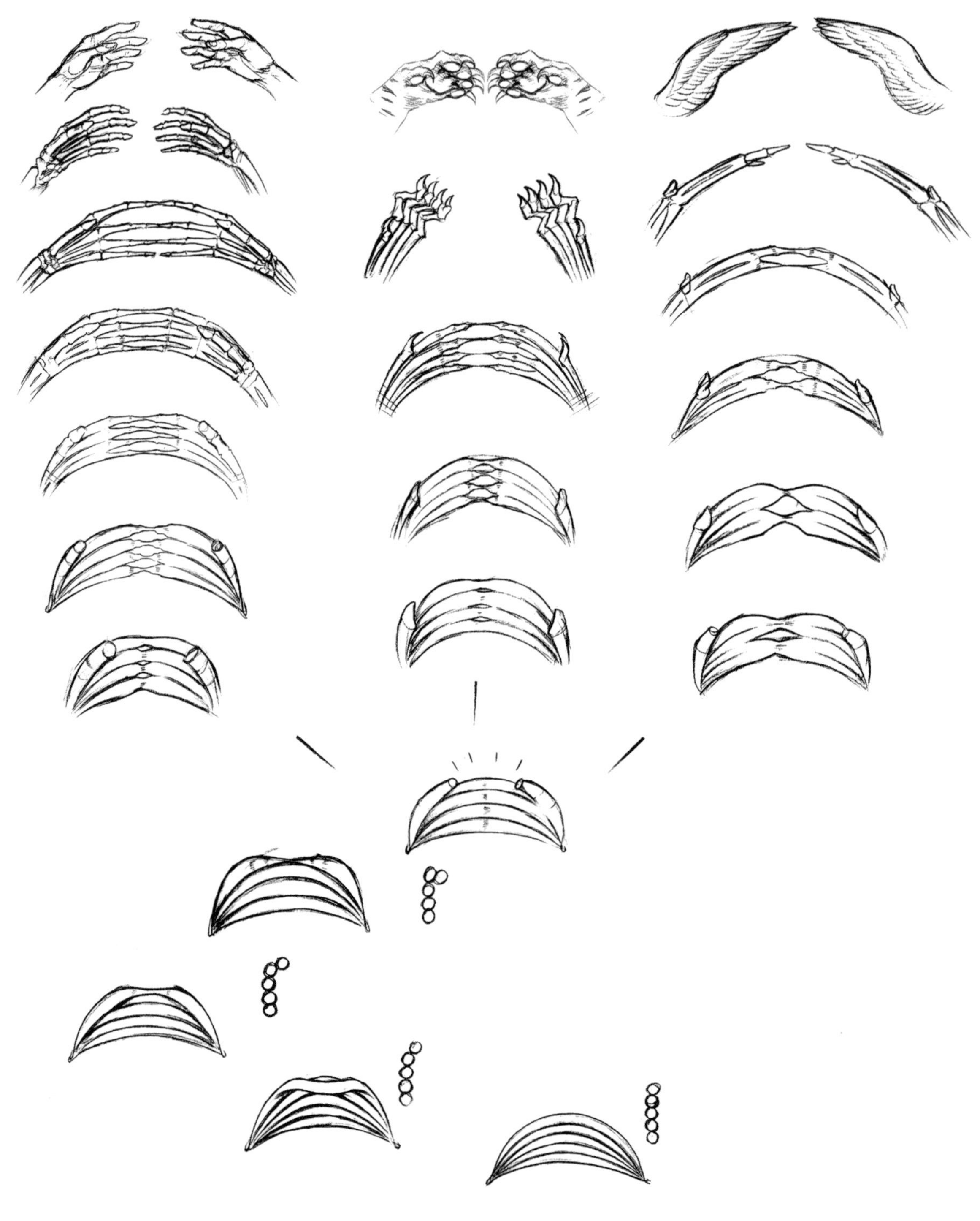

Human, Feline, and Avian Limb Development

Plate 7

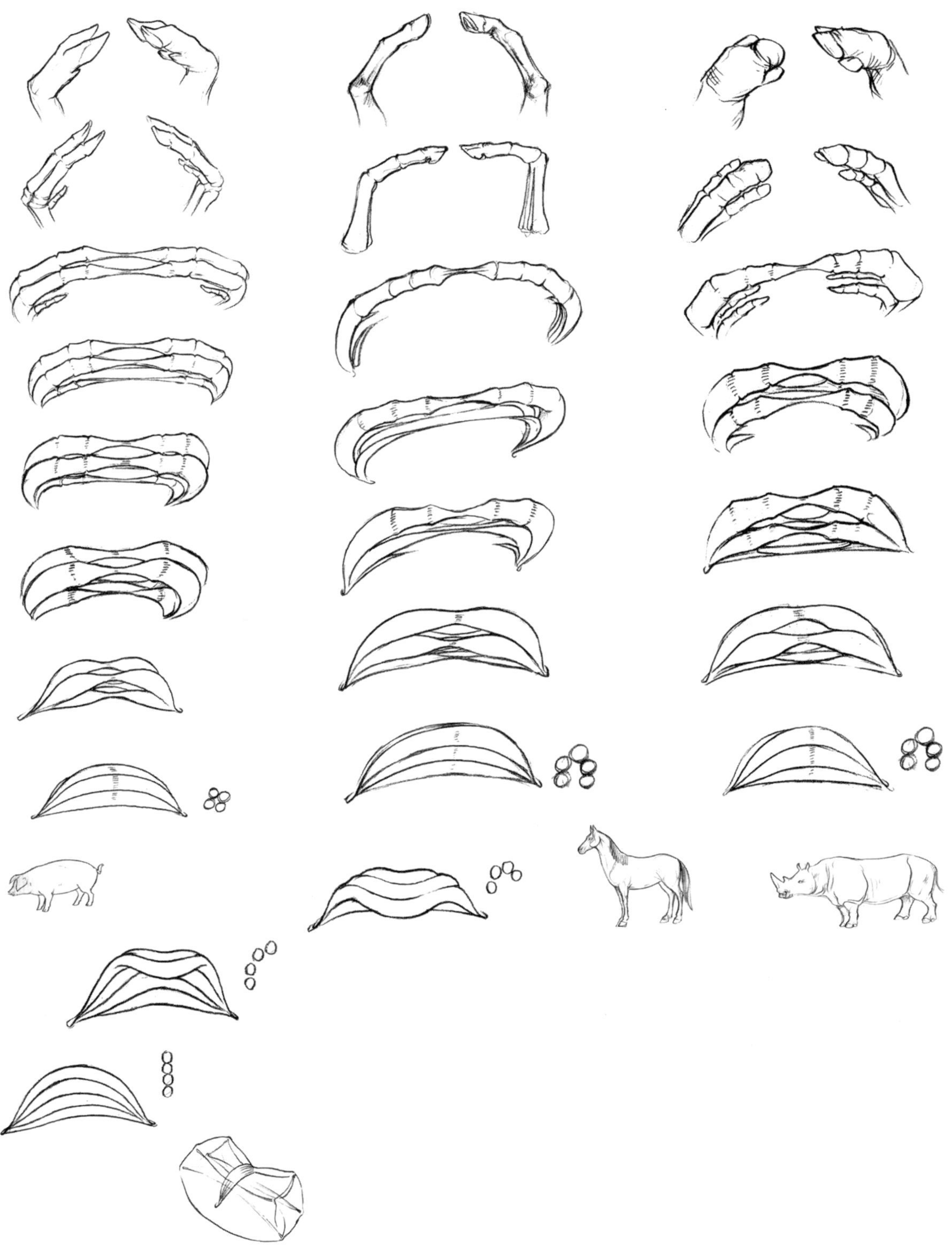

Pig, Horse, and Rhinoceros Limb Development

Plate 8

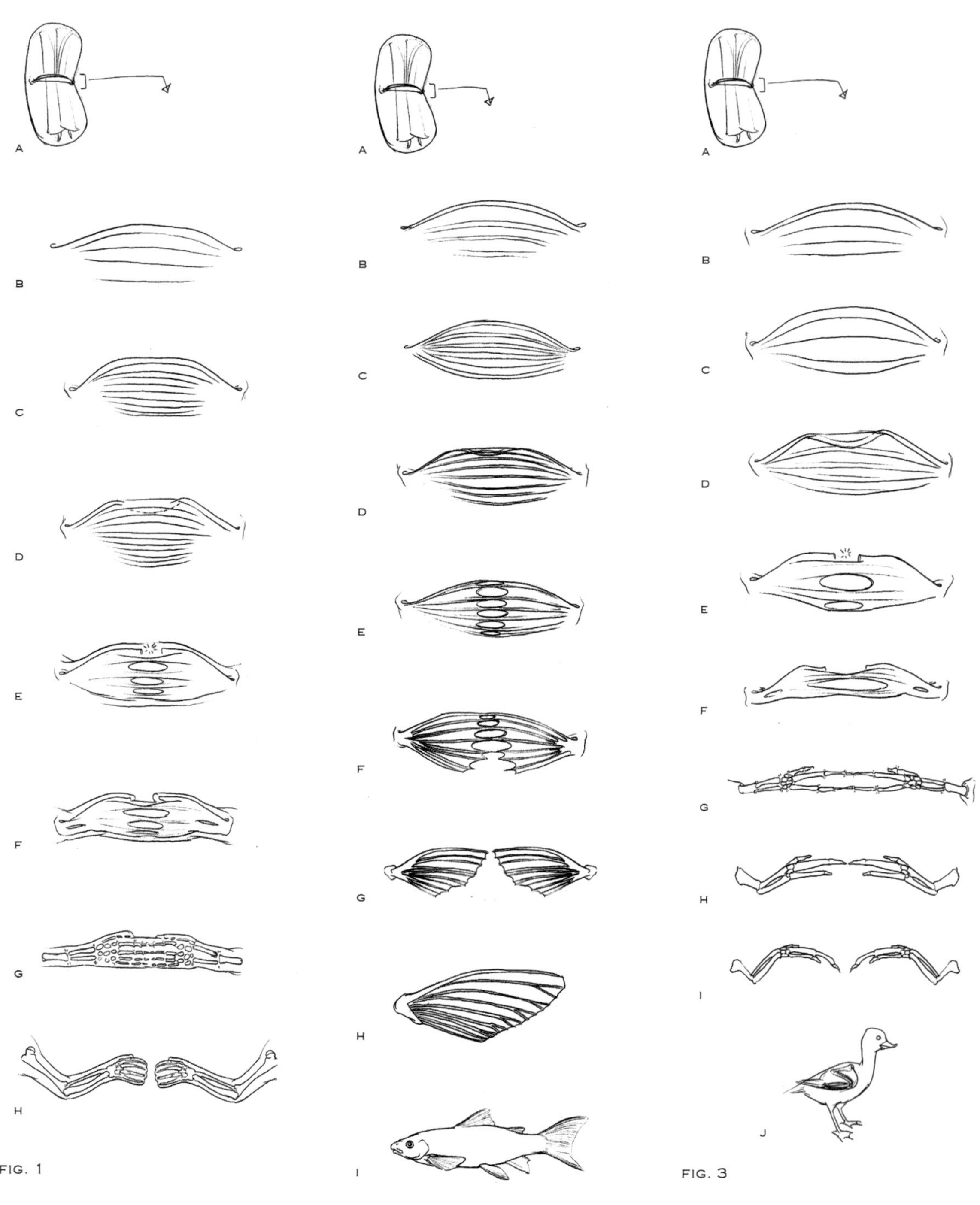

LIMB DEVELOPMENT BY MORPHOGENETIC FIELDS

PLATE 9

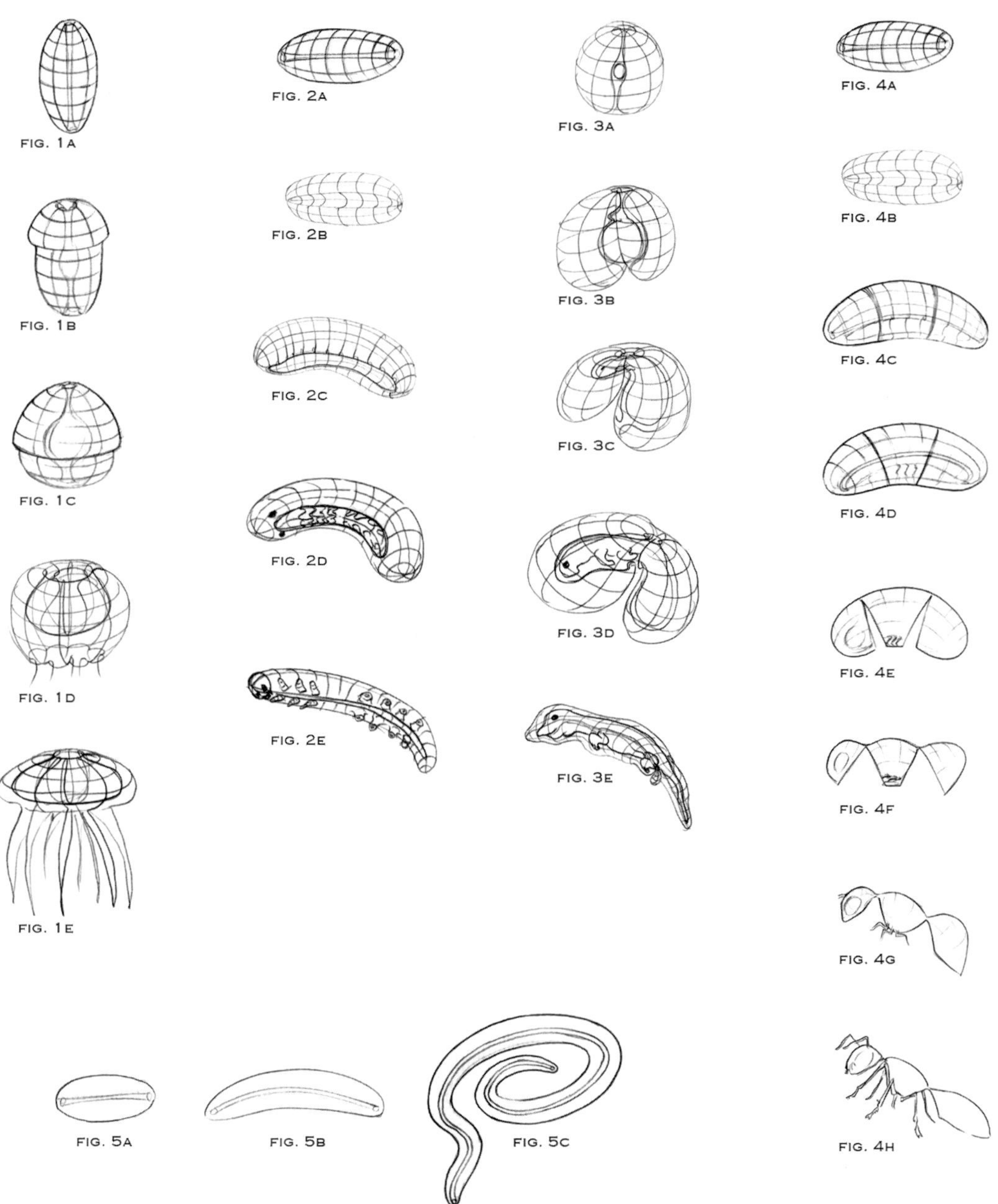

The Common Geometric Origin of Various Phyla

Plate 10

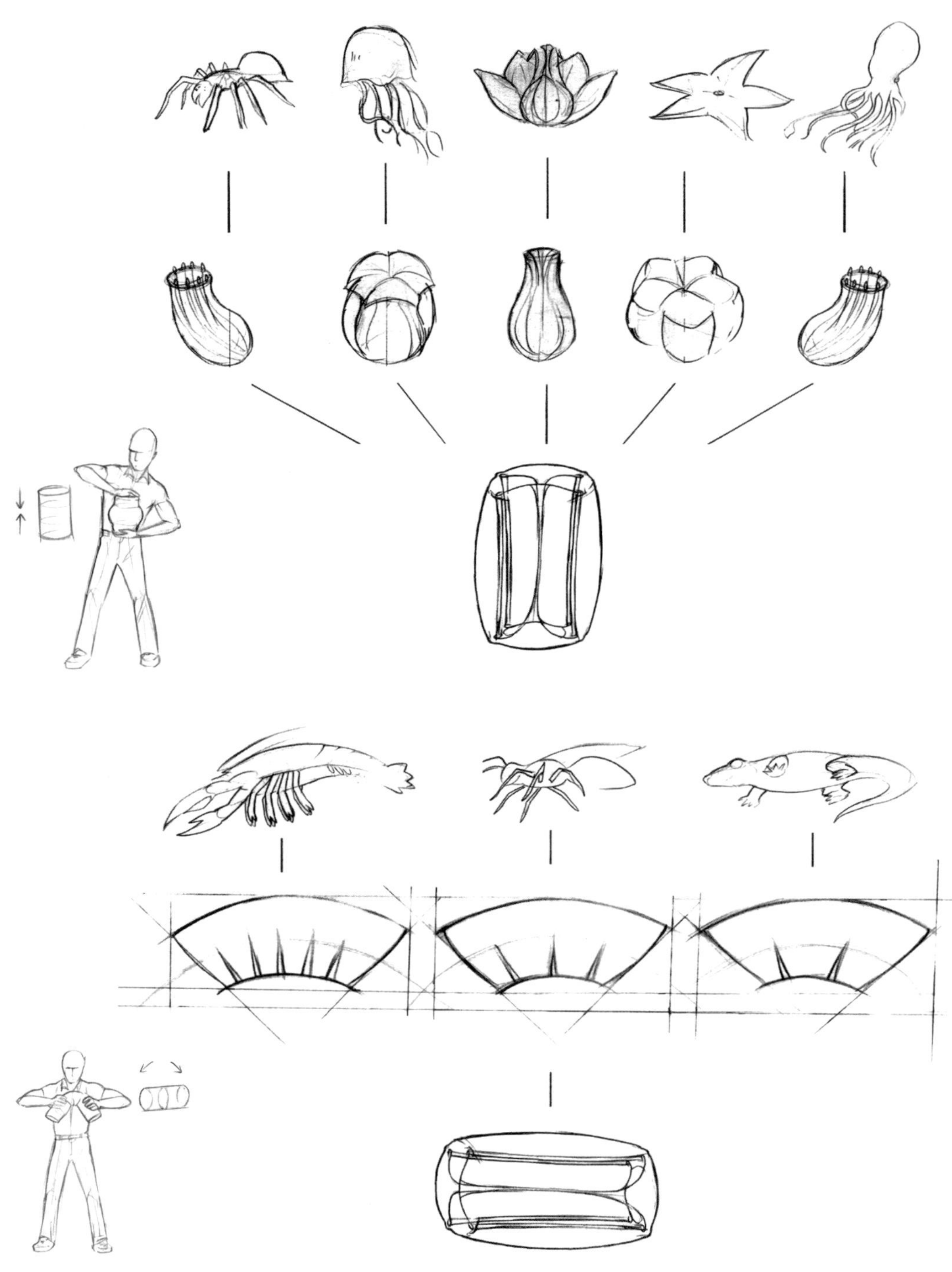

The Origin of Phyletic Form by Mechanical Deformation

Plate 11

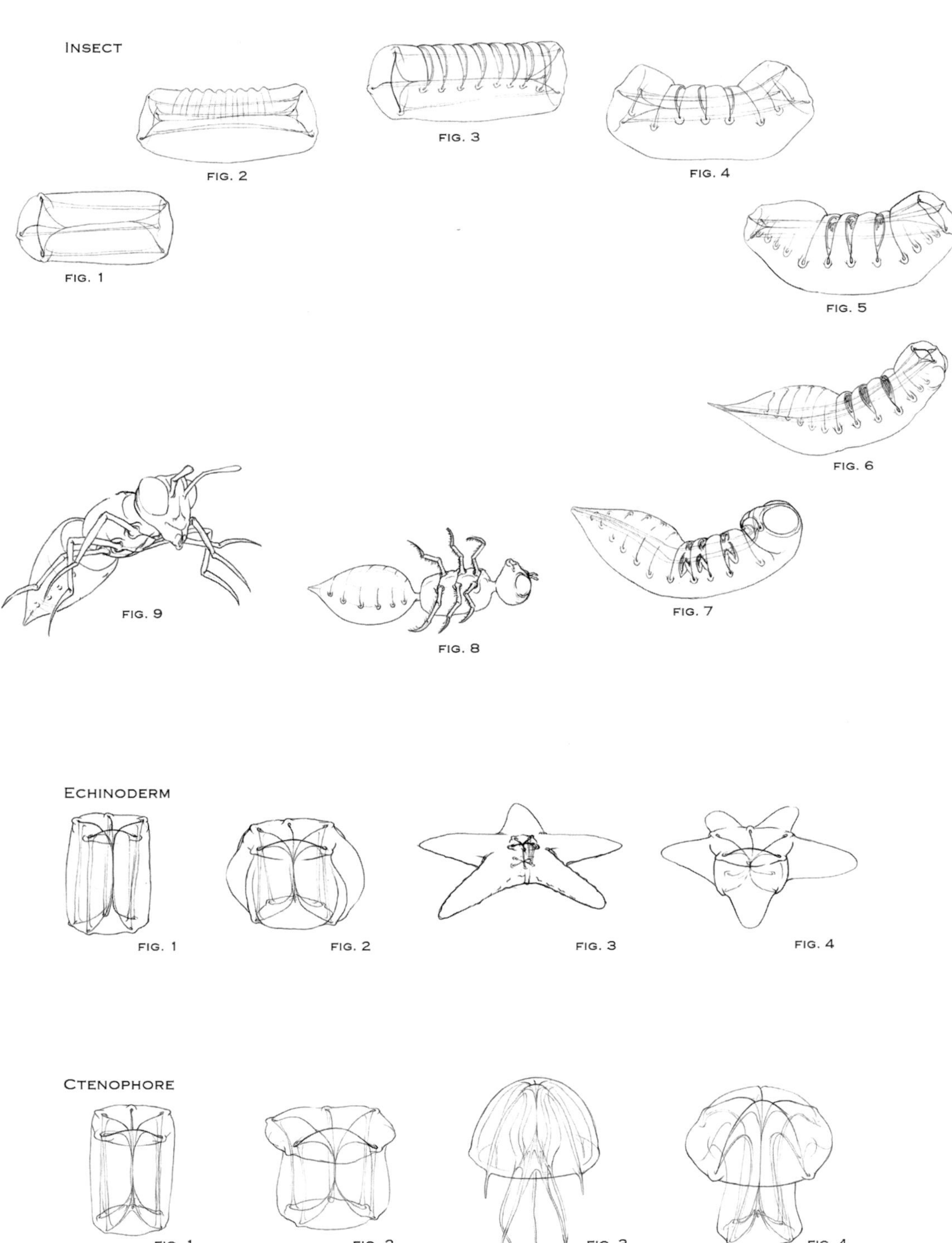

Development of Insect, Echinoderm, and Ctenophore

Plate 12

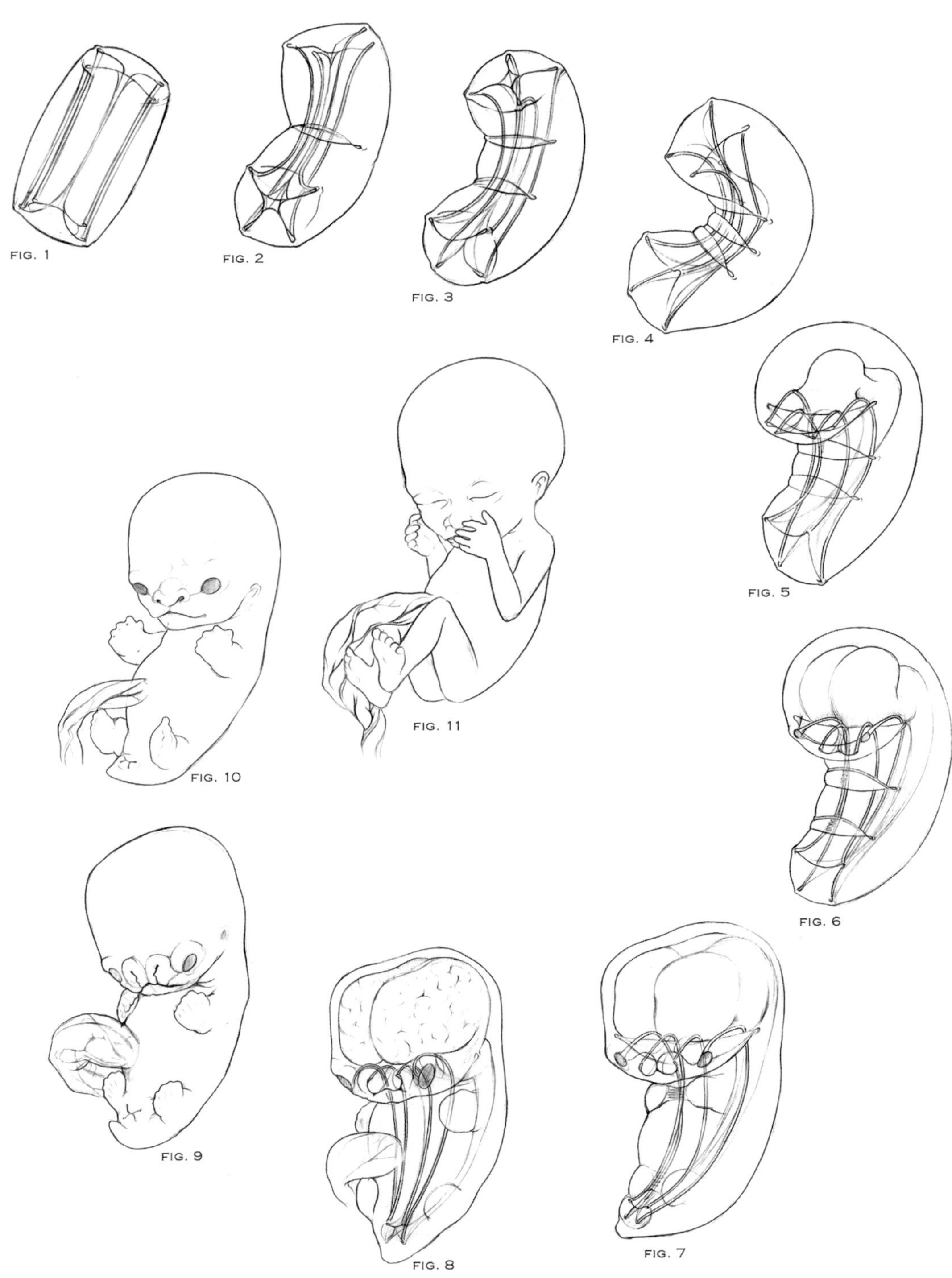

VERTEBRATE DEVELOPMENT

PLATE 13

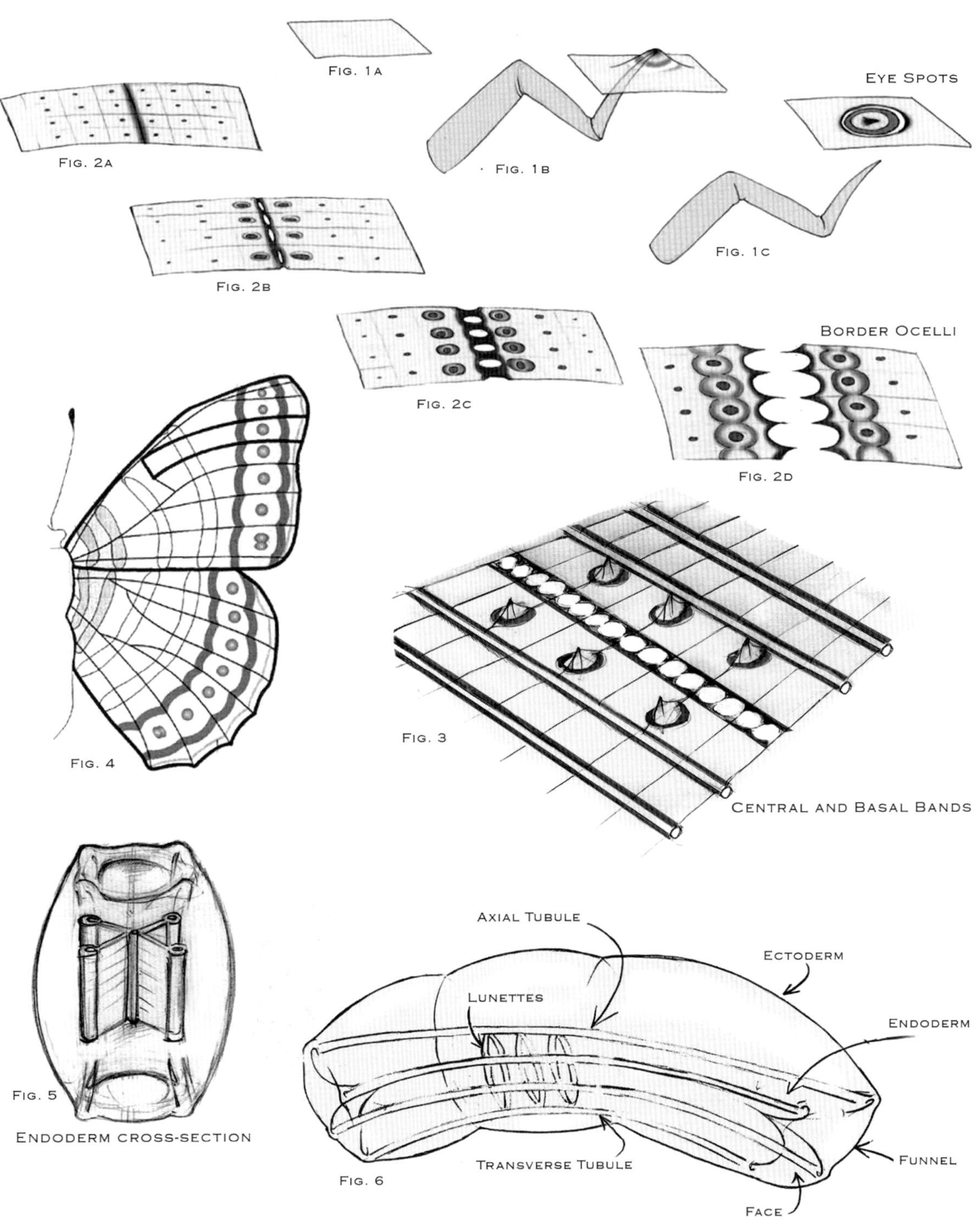

Mechanical Origin of Butterfly Wing Pattern

Plate 14

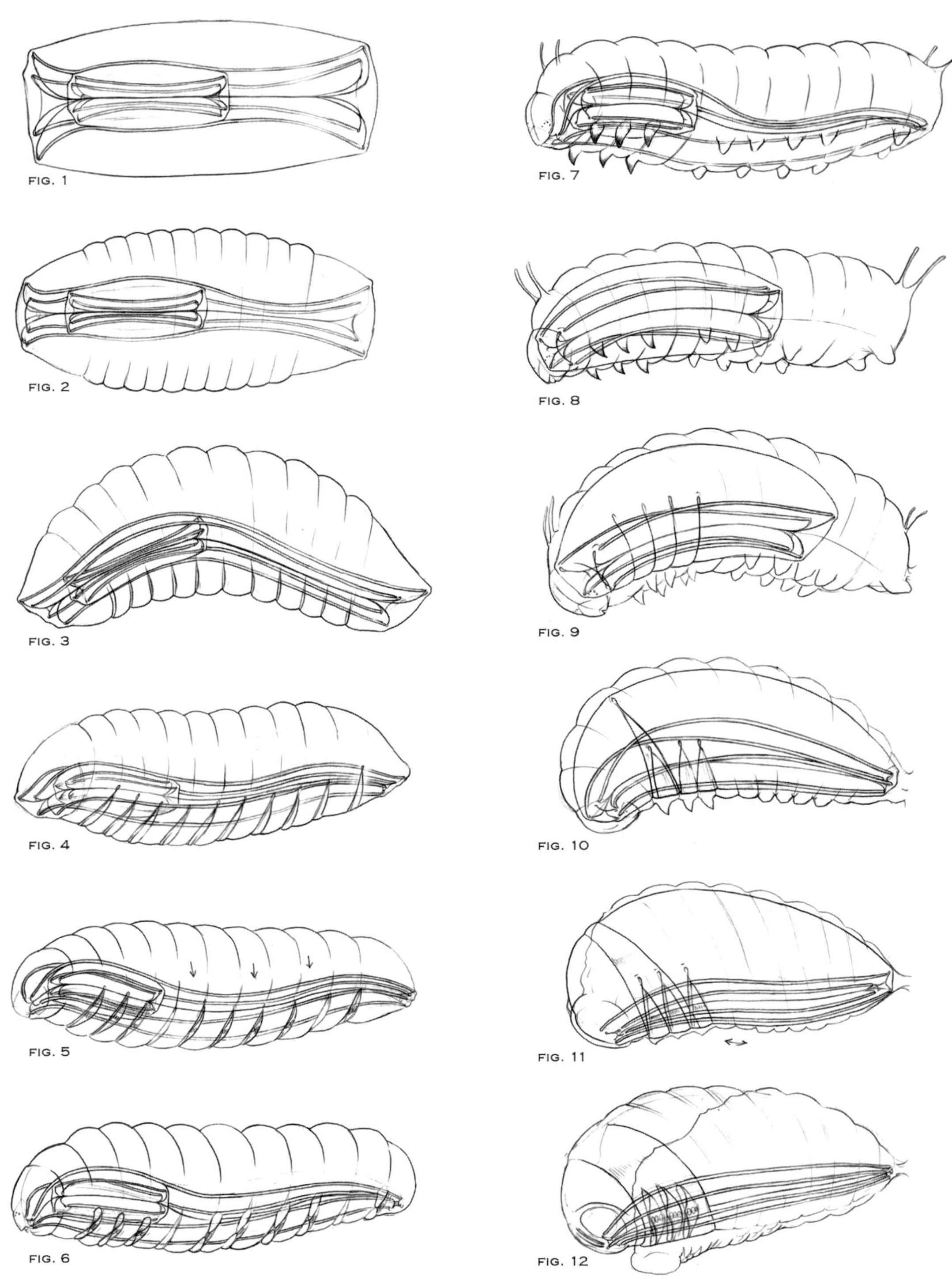

BUTTERFLY METAMORPHOSIS

PLATE 15A

PLATE 15B

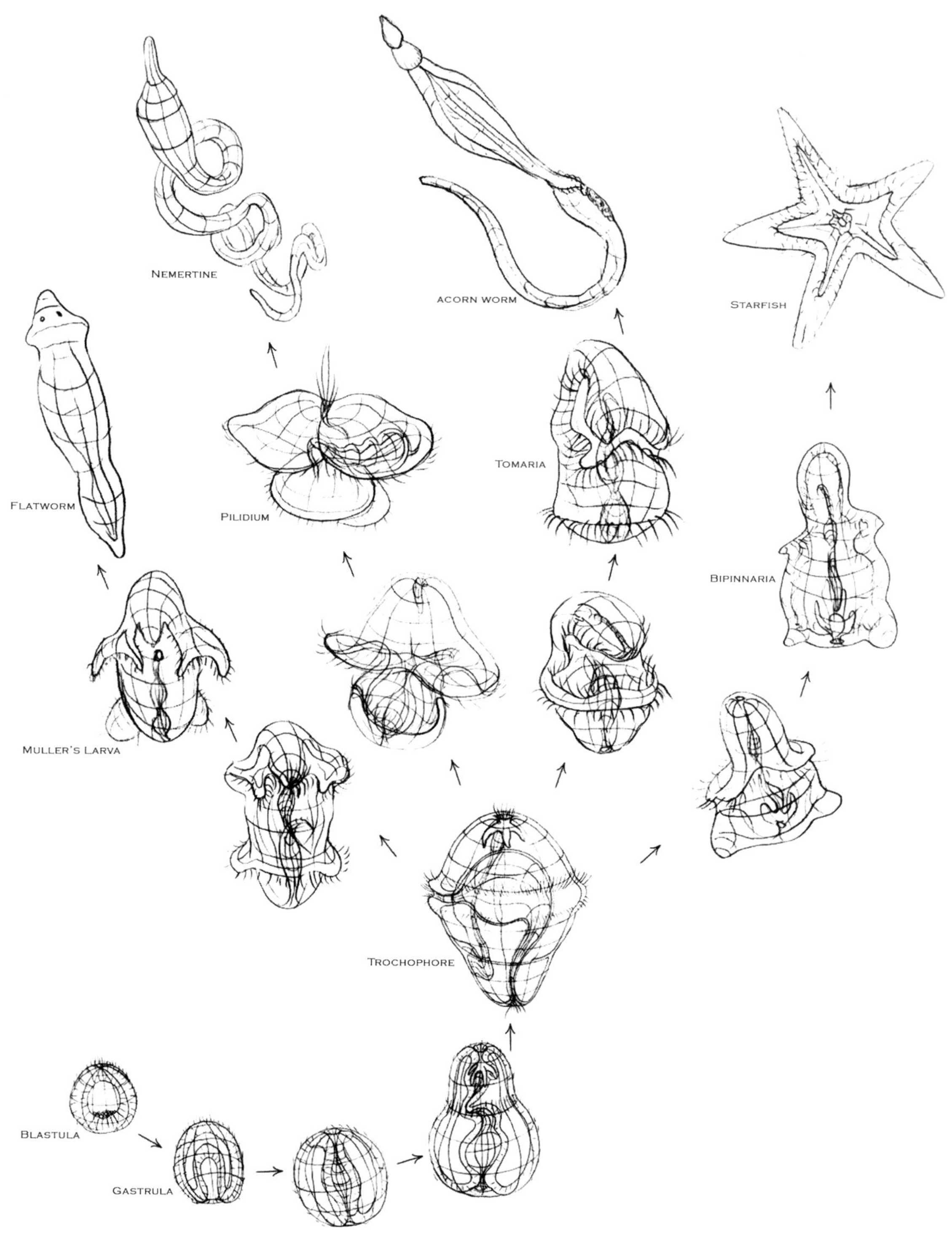

Larval Development

Plate 16

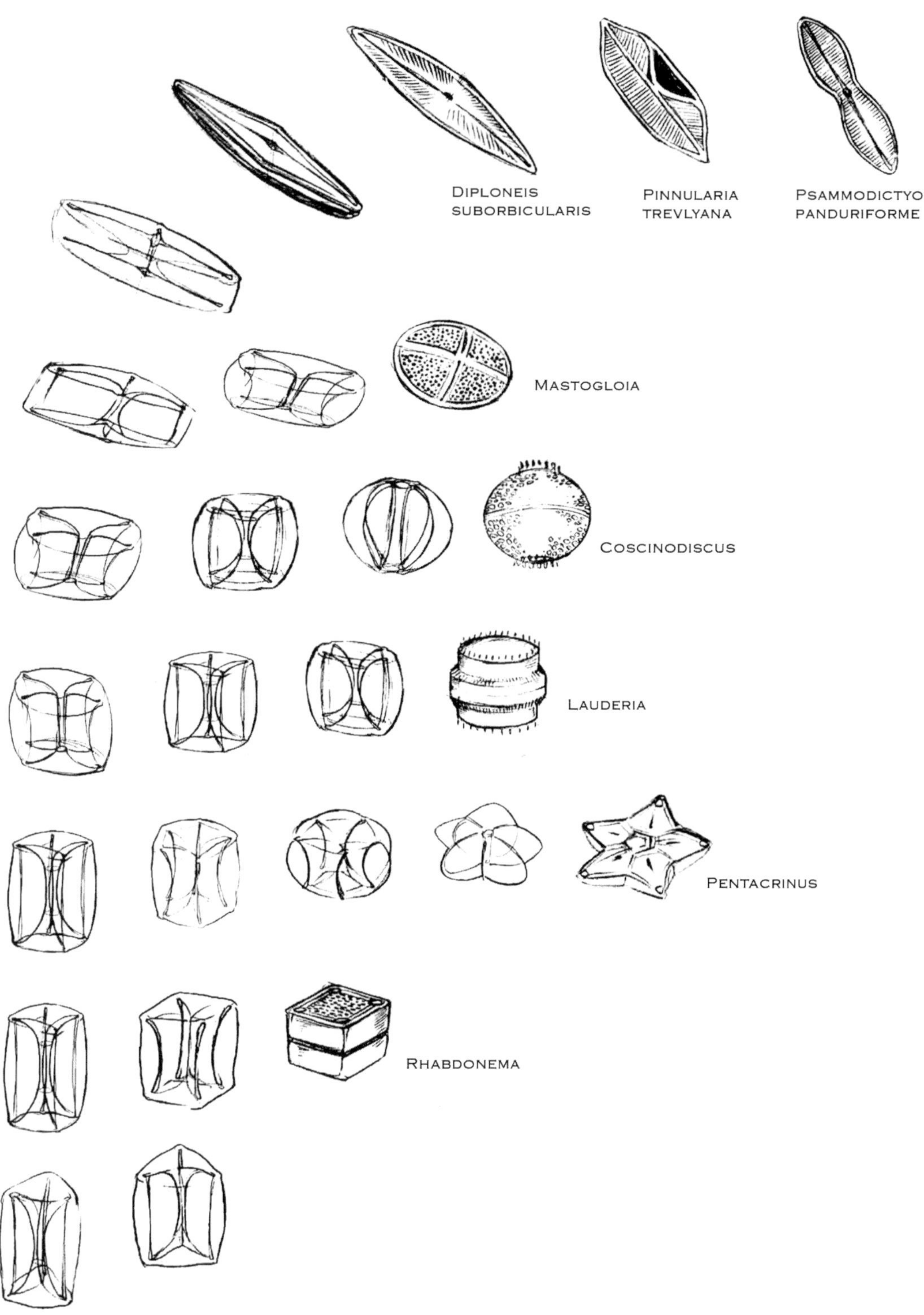

Schematic Origin of Diatom Forms

Plate 17

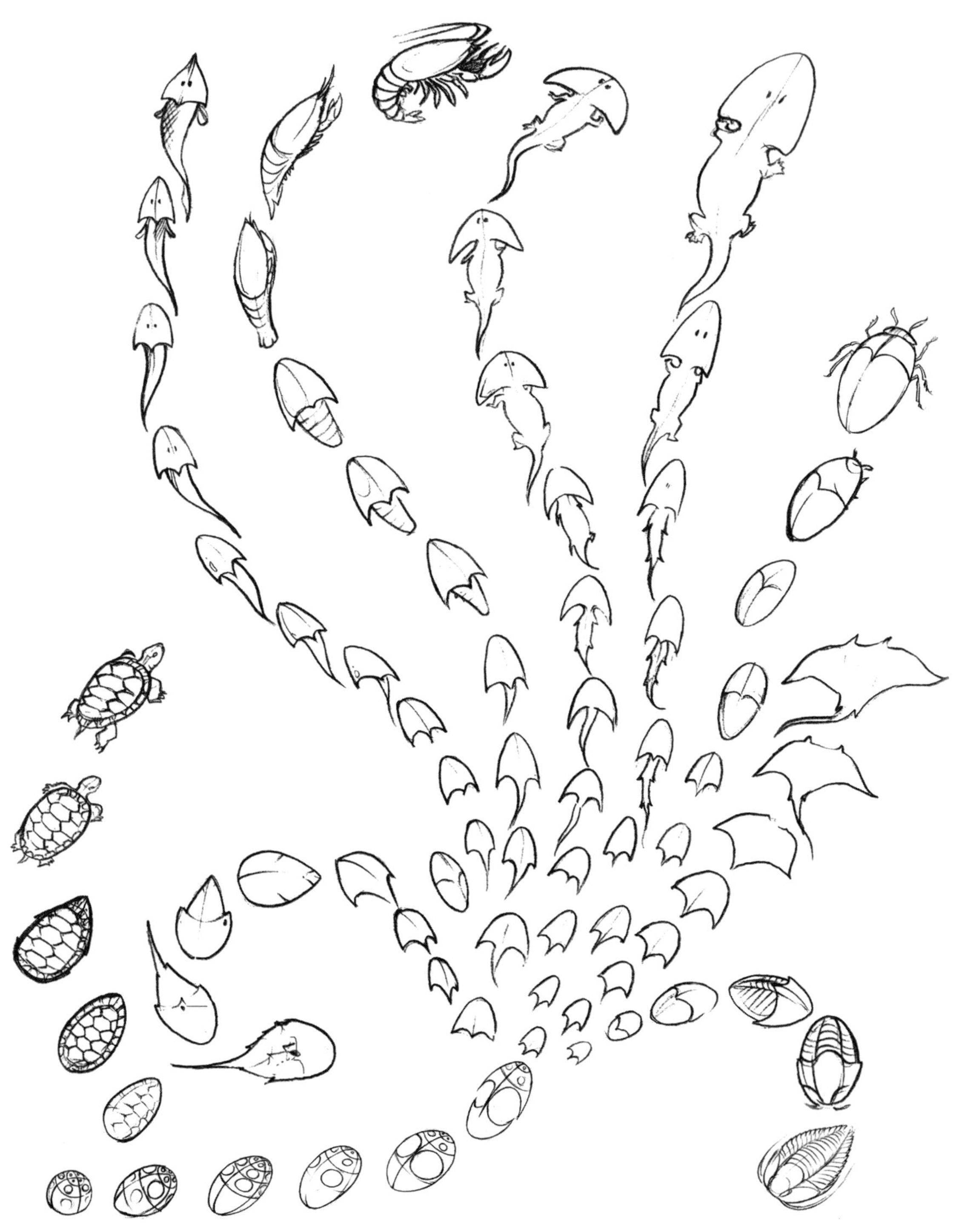

Schematic Origin of Bilateral Bio-Diversity

Plate 18

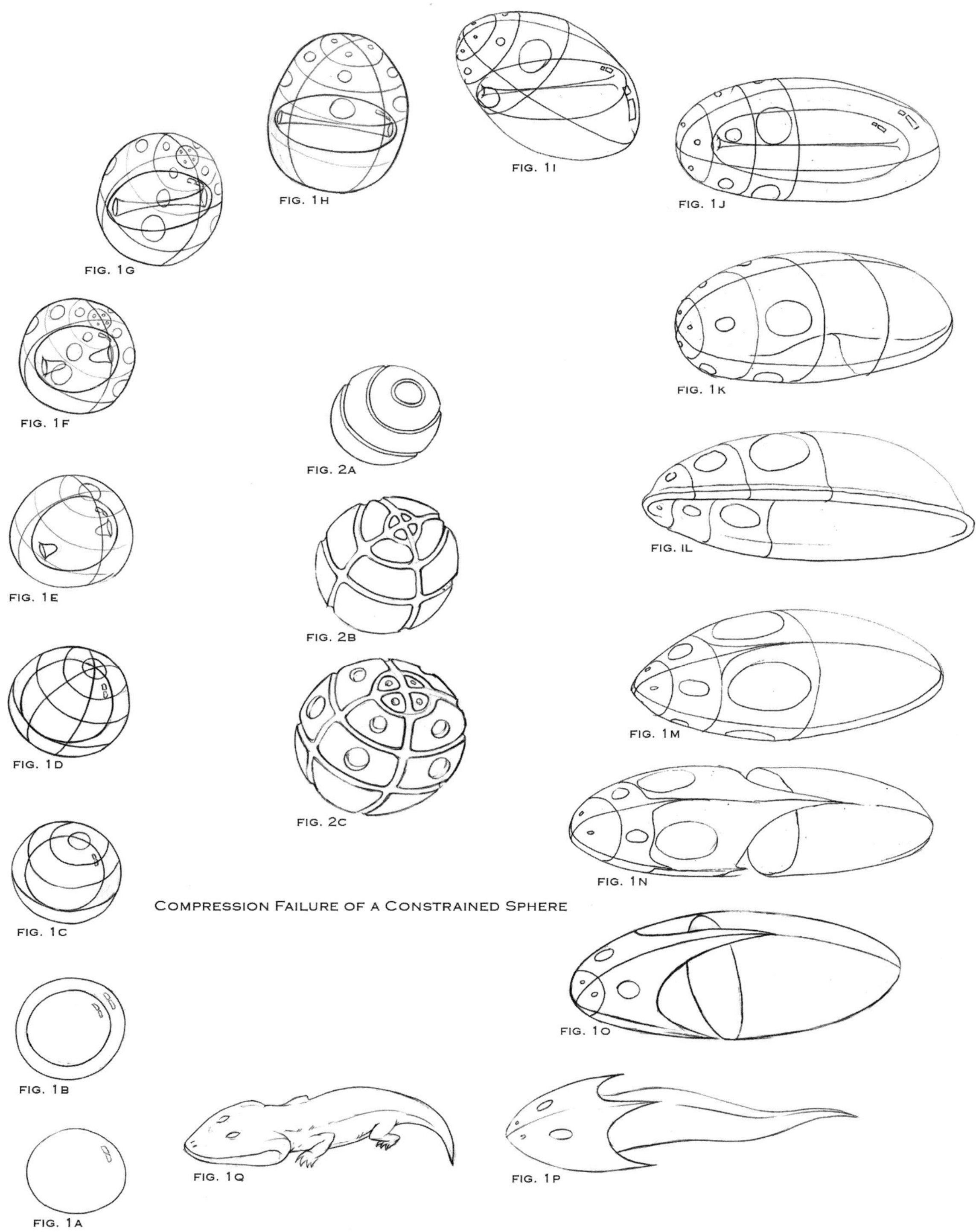

Schematic Origin of the Bilateral Body Form

Plate 19

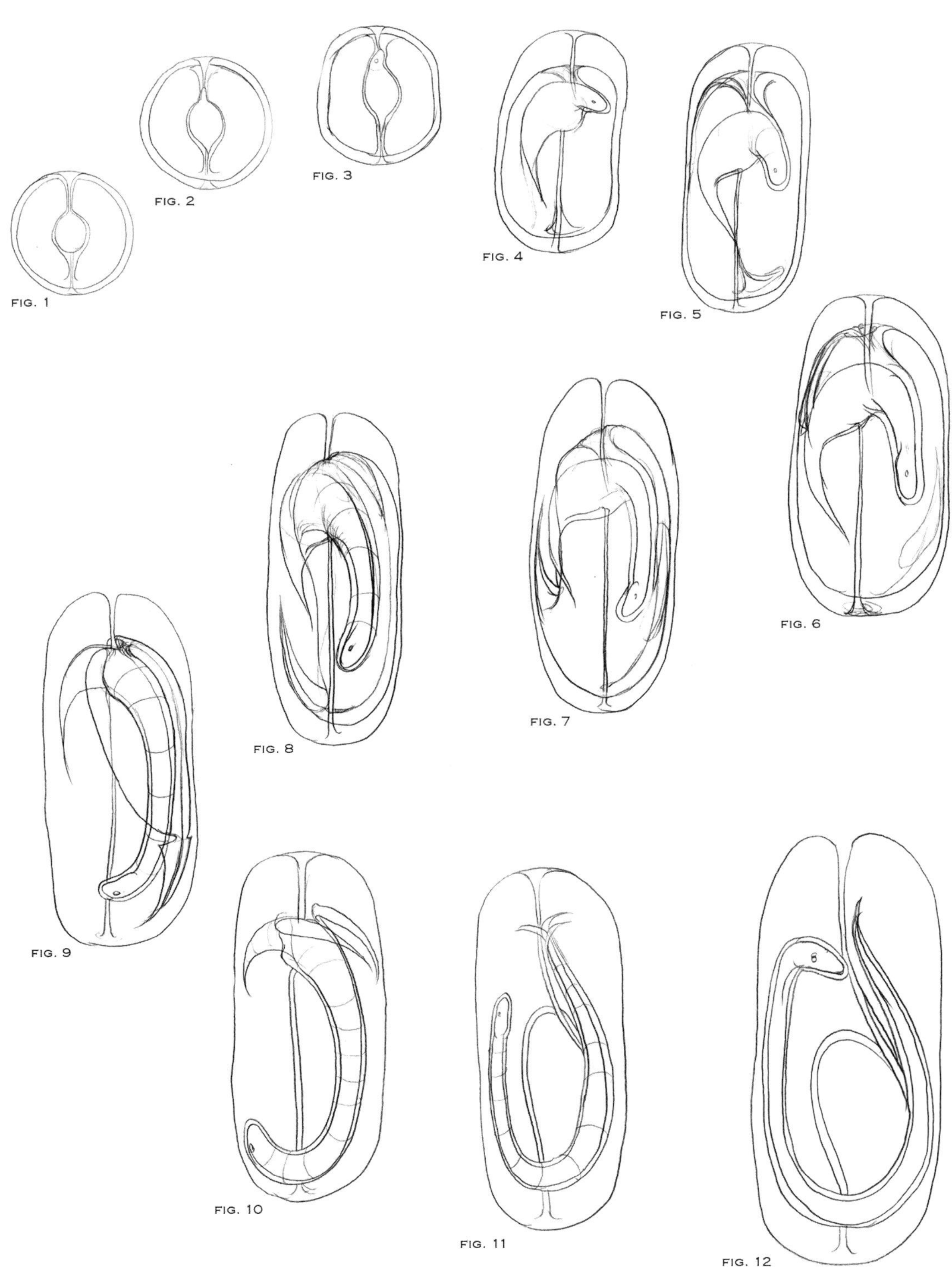

Vertebrate Development

Plate 20

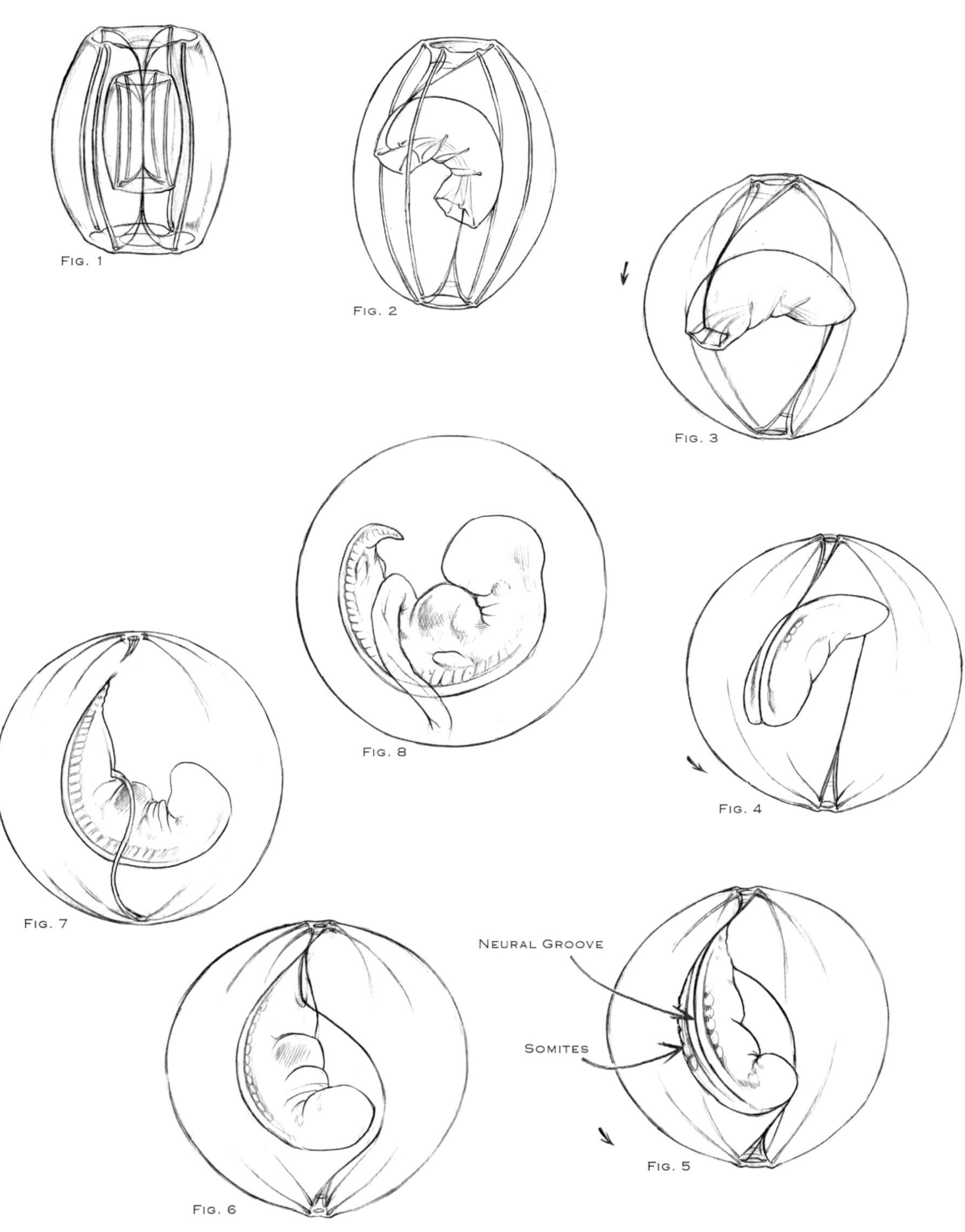

Establishment of the Vertebrate Neural System

Plate 21

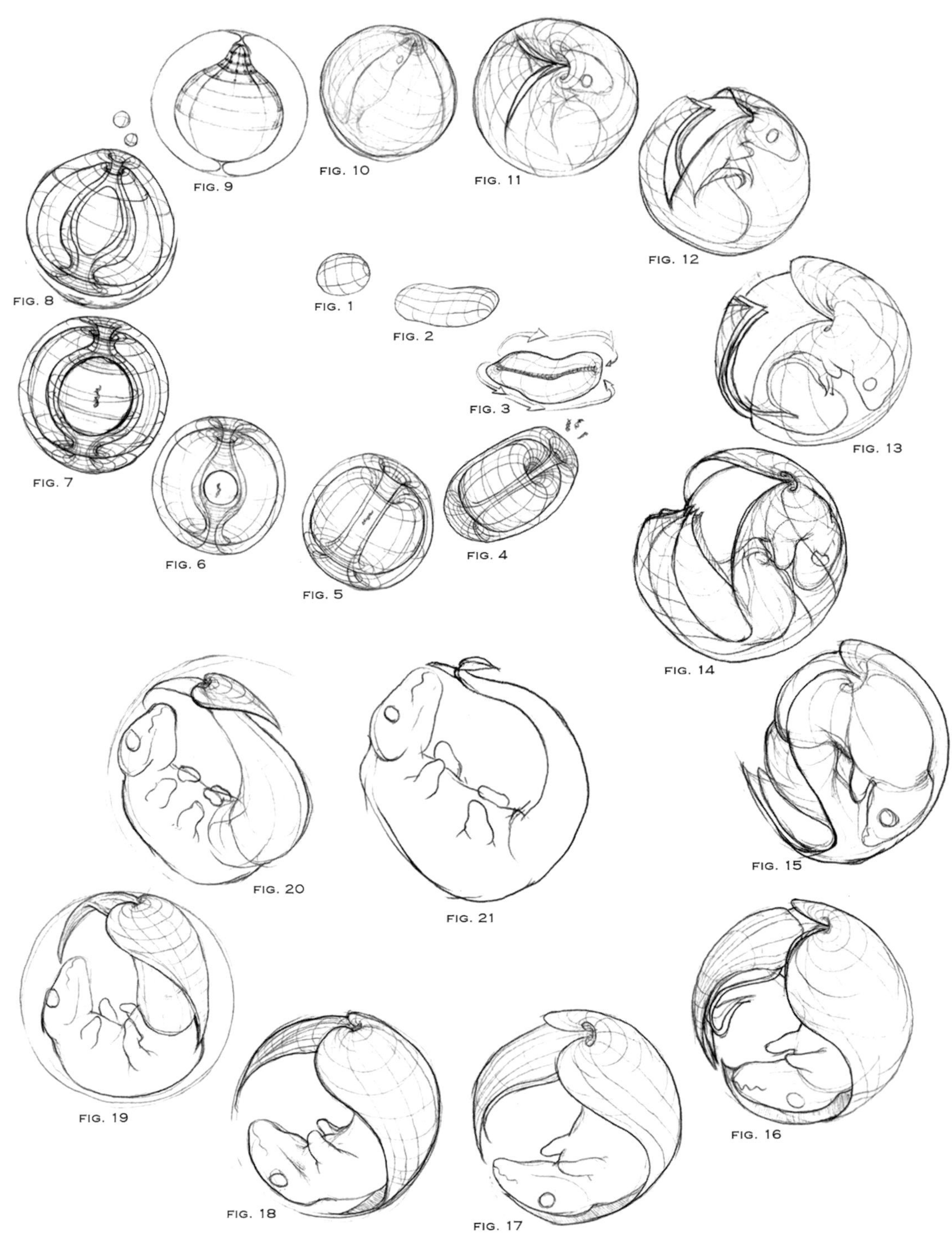

Vertebrate Development

Plate 22

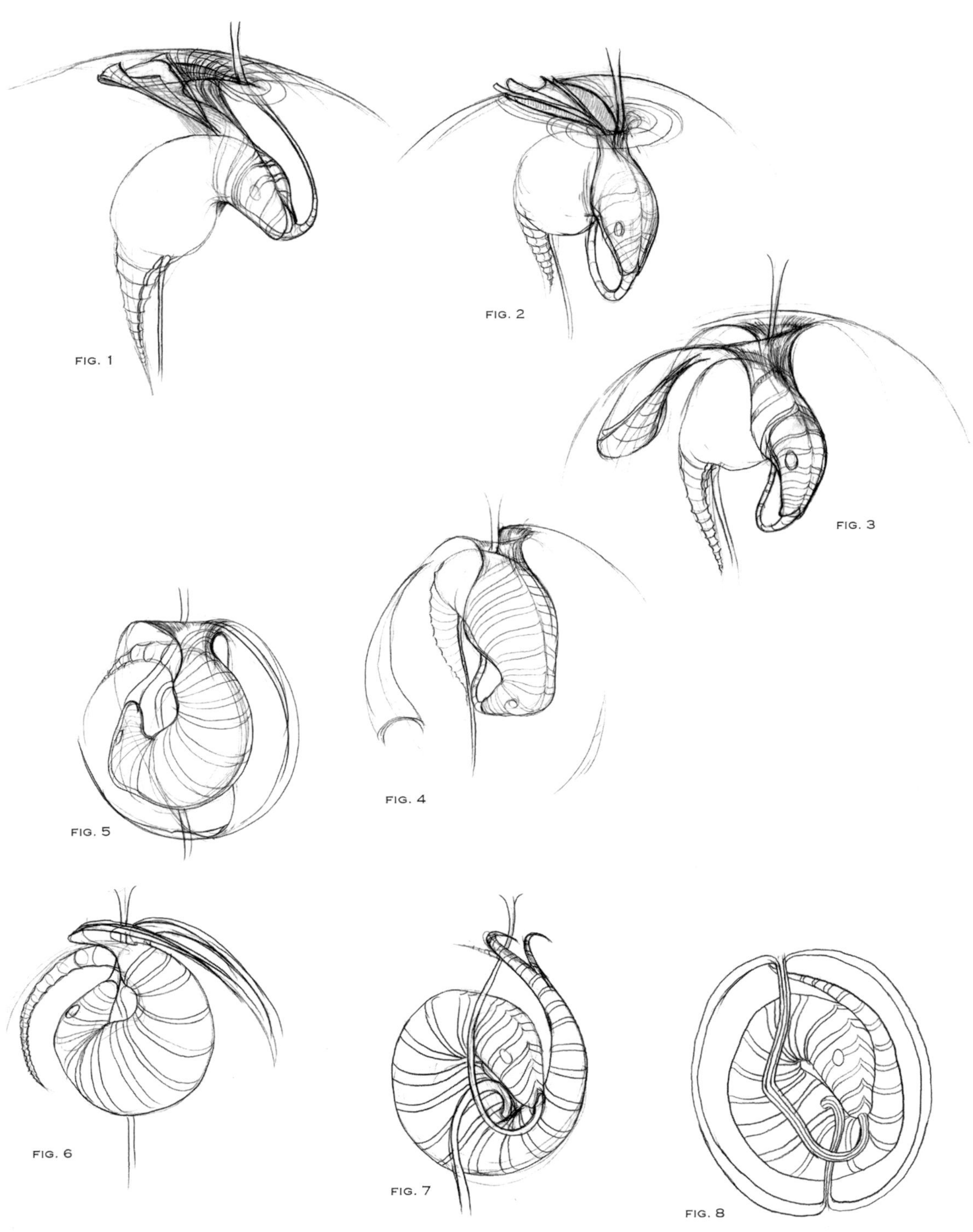

Vertebrate Development

Plate 23

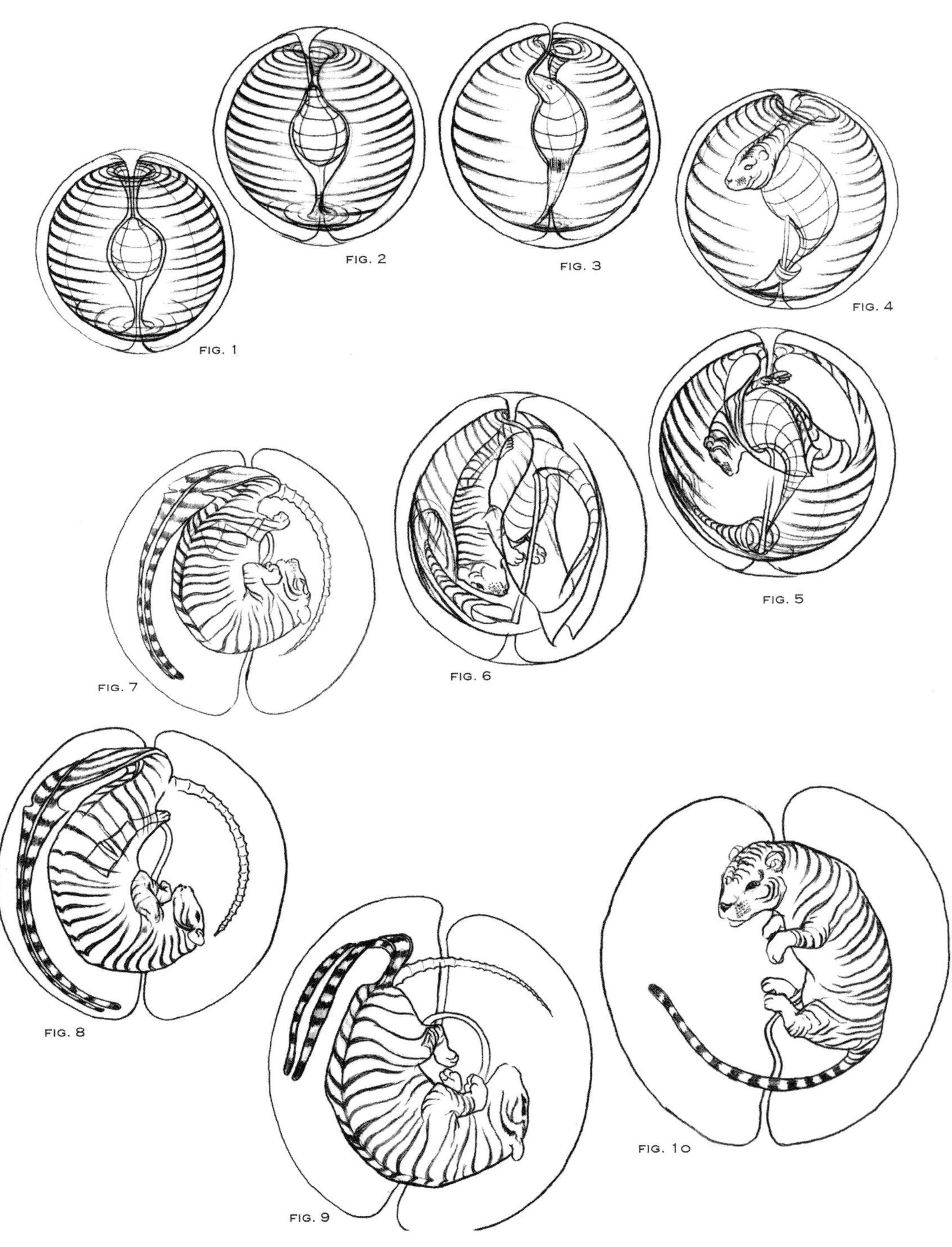

Stripe Pattern Development

Plate 24

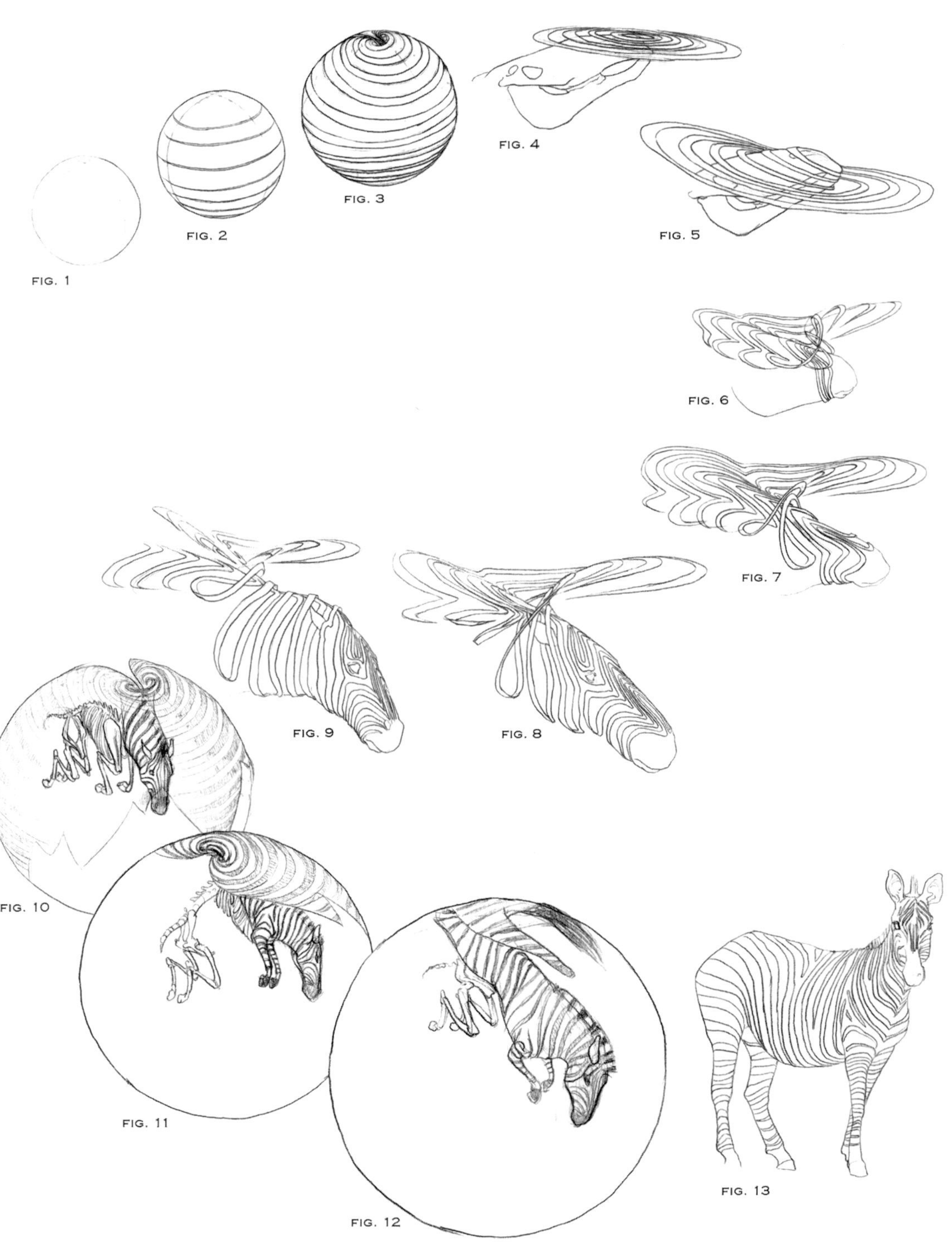

Stripe Pattern Development

Plate 25

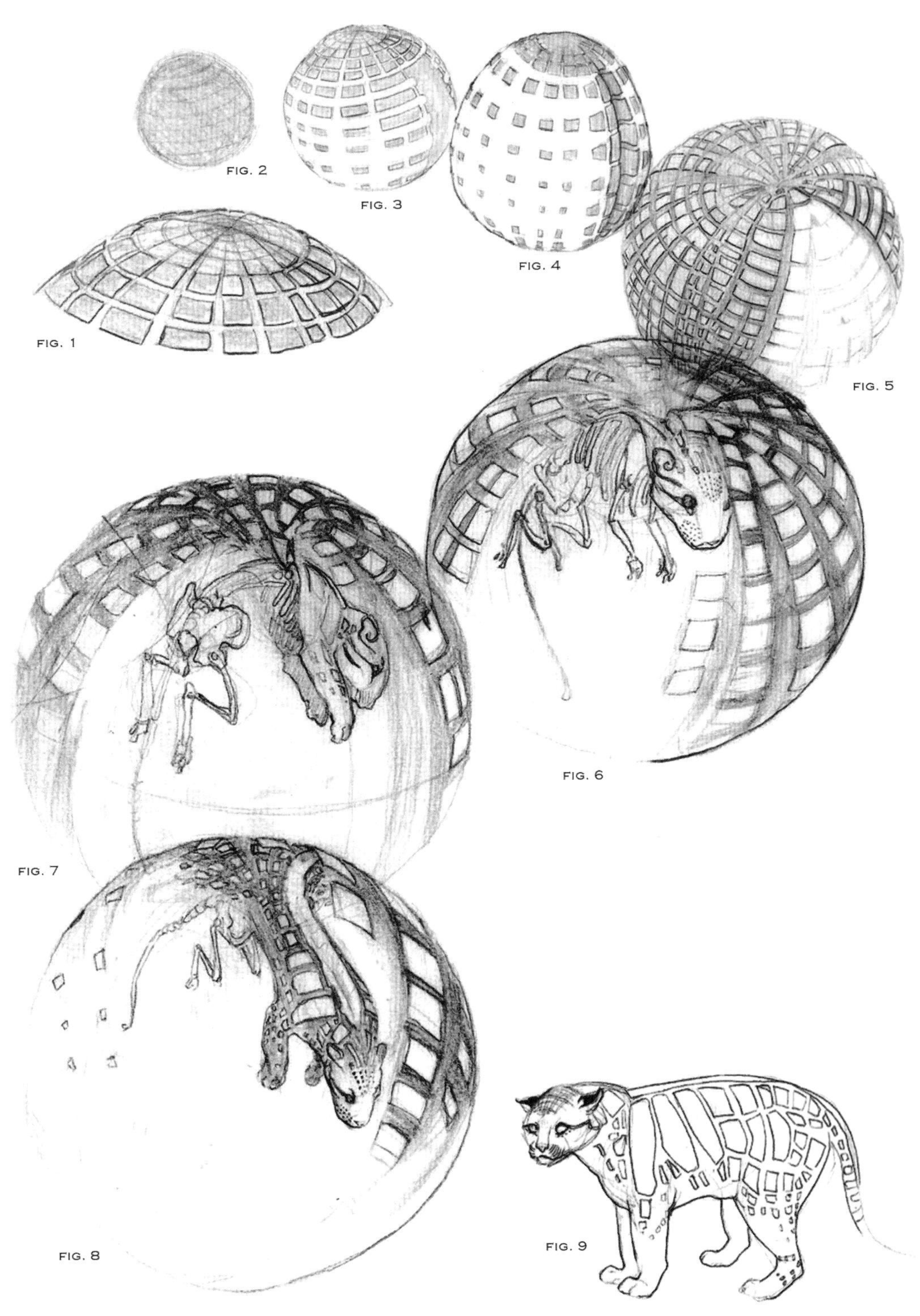

Clouded Leopard Development

Plate 26

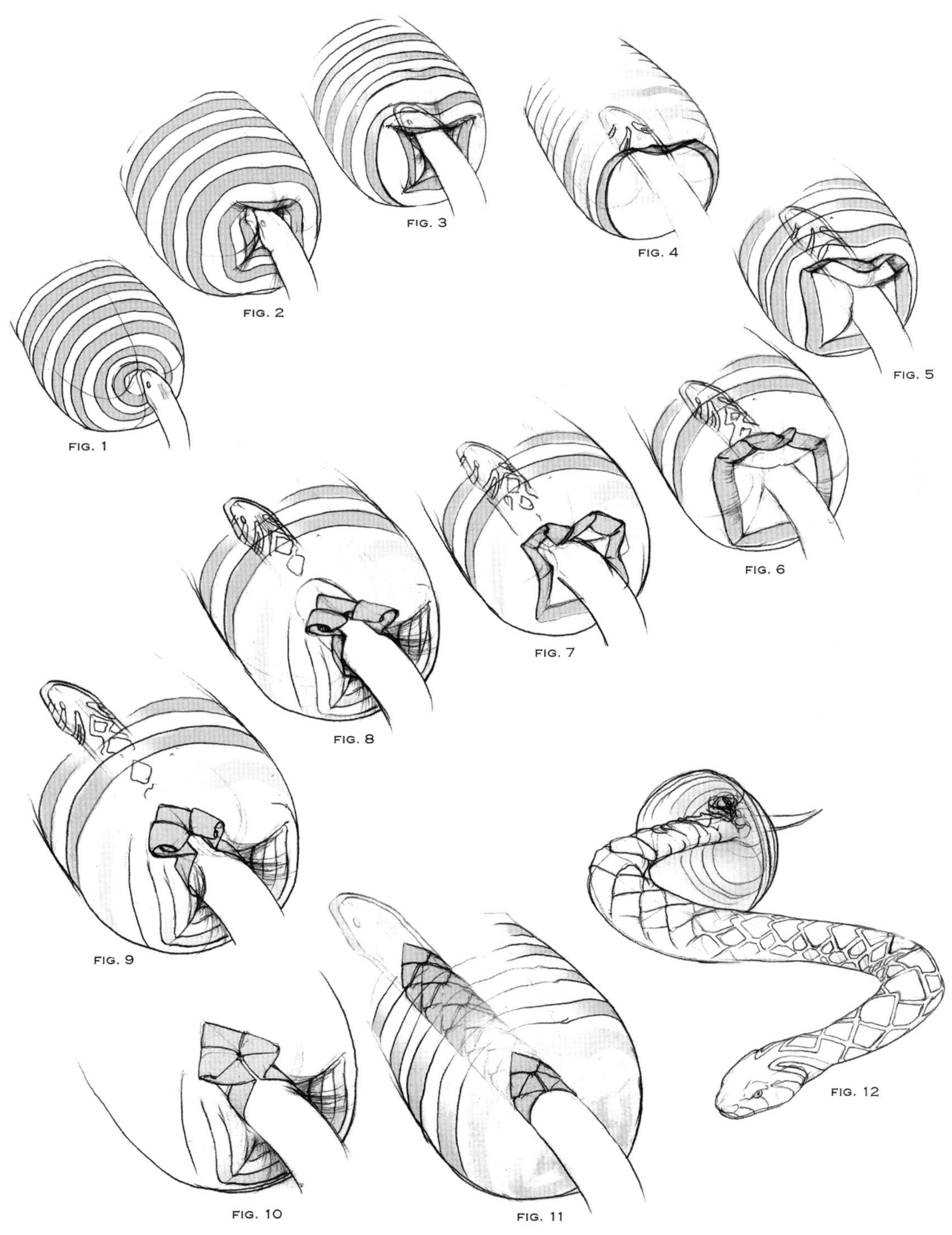

Snake Skin Pattern Development

Plate 27

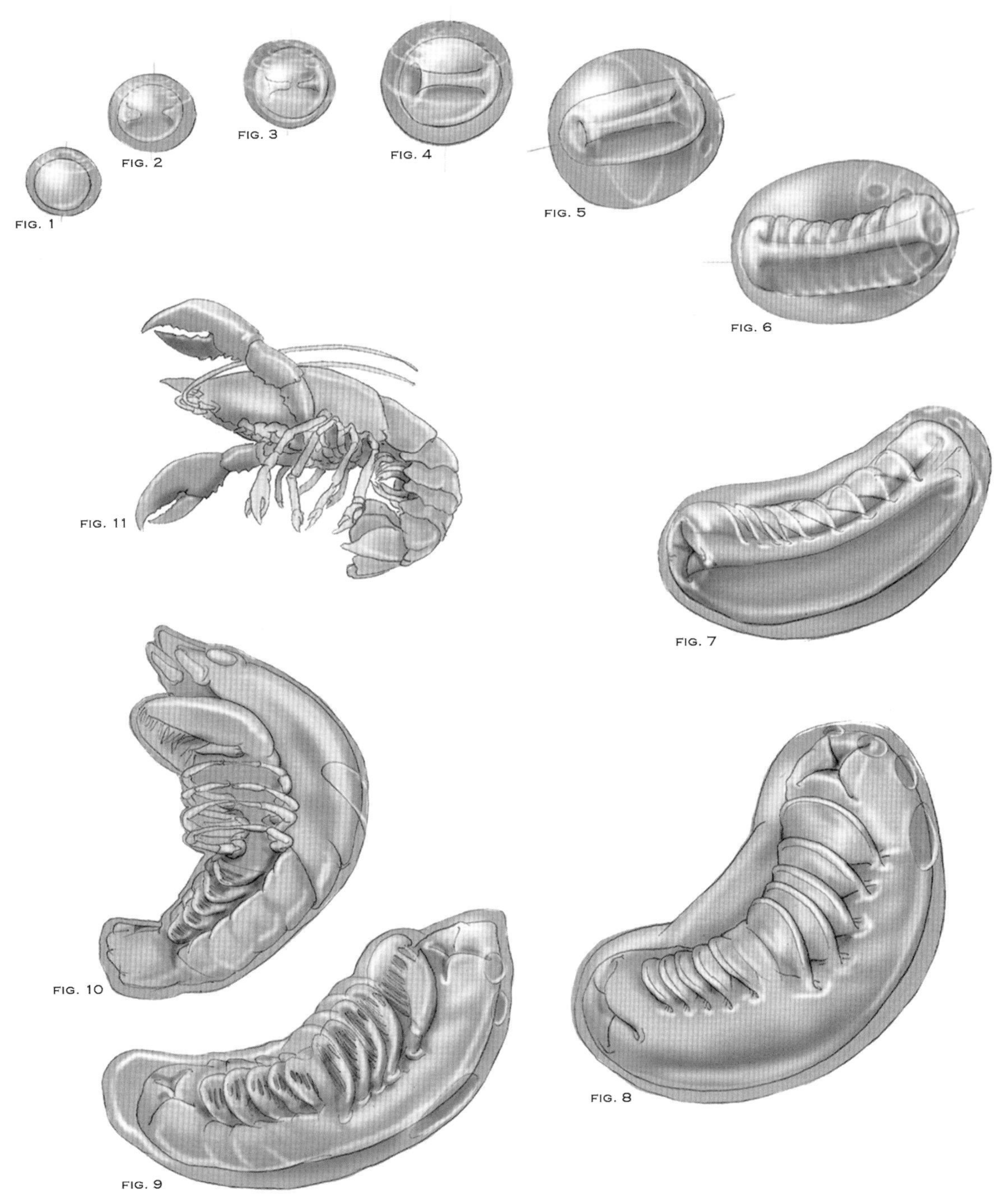

Crustacean Development

Plate 28

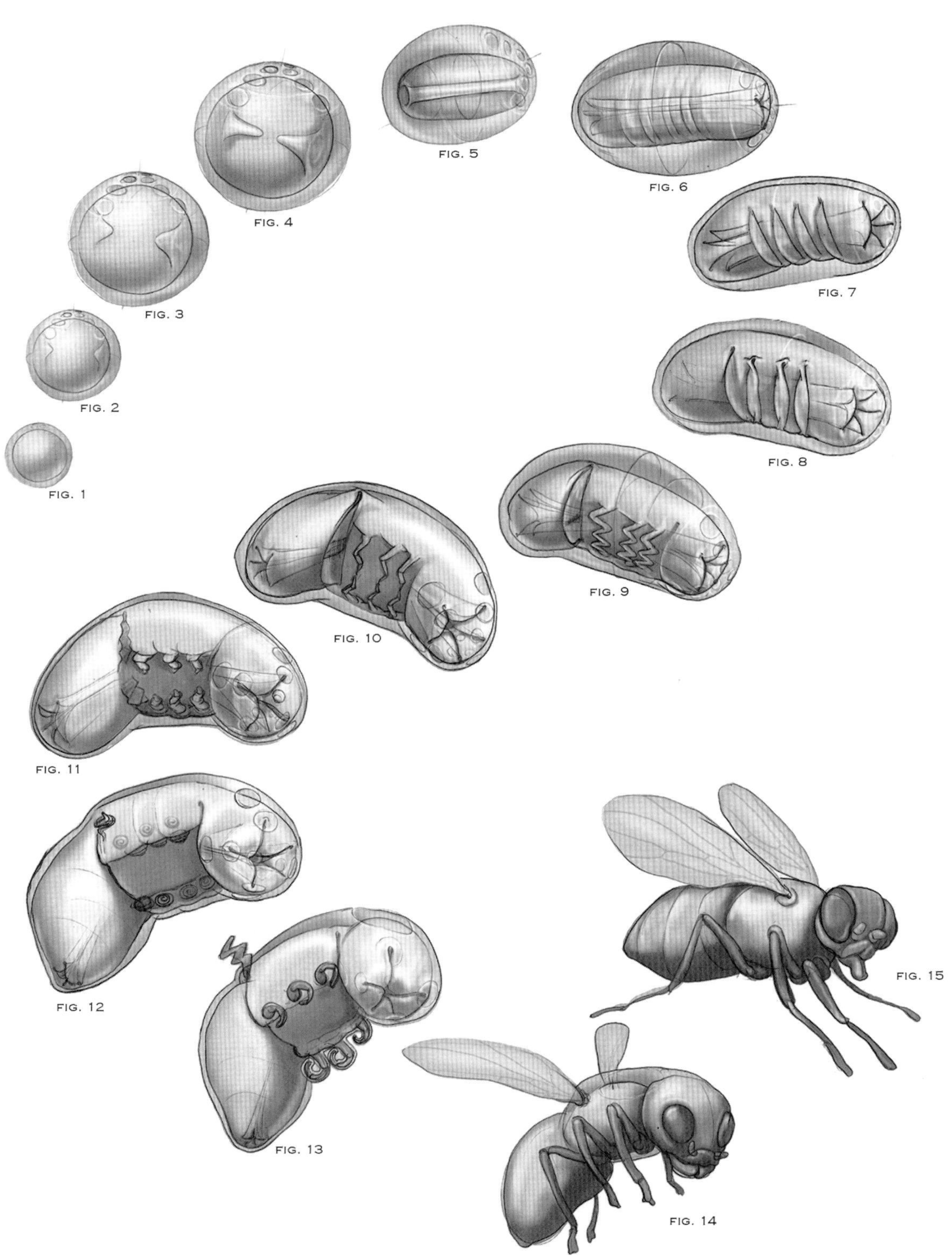

Fly Development by Morphogenetic Fields

Plate 29

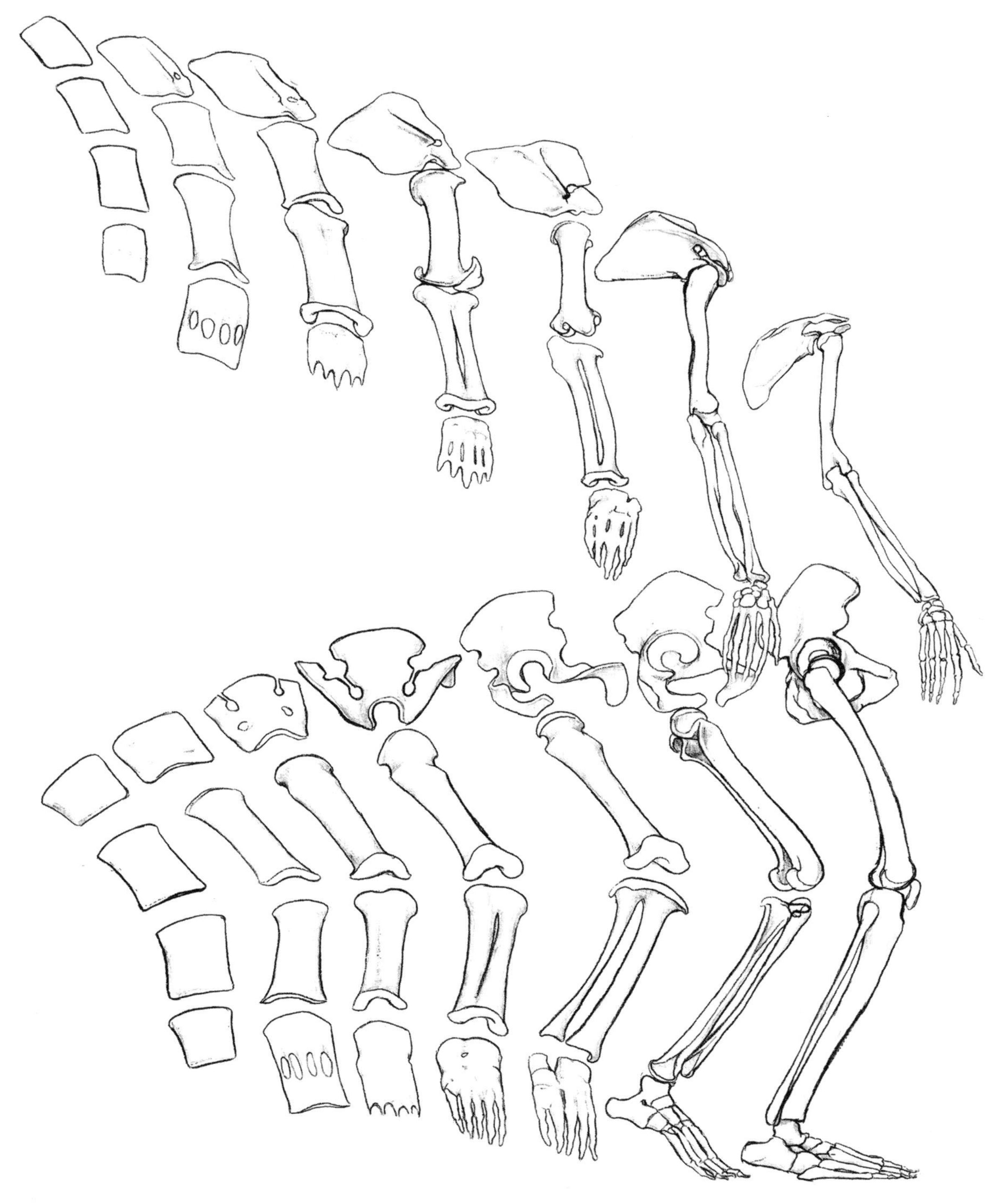

Morphogenesis of the Human Skeleton
Plate 30

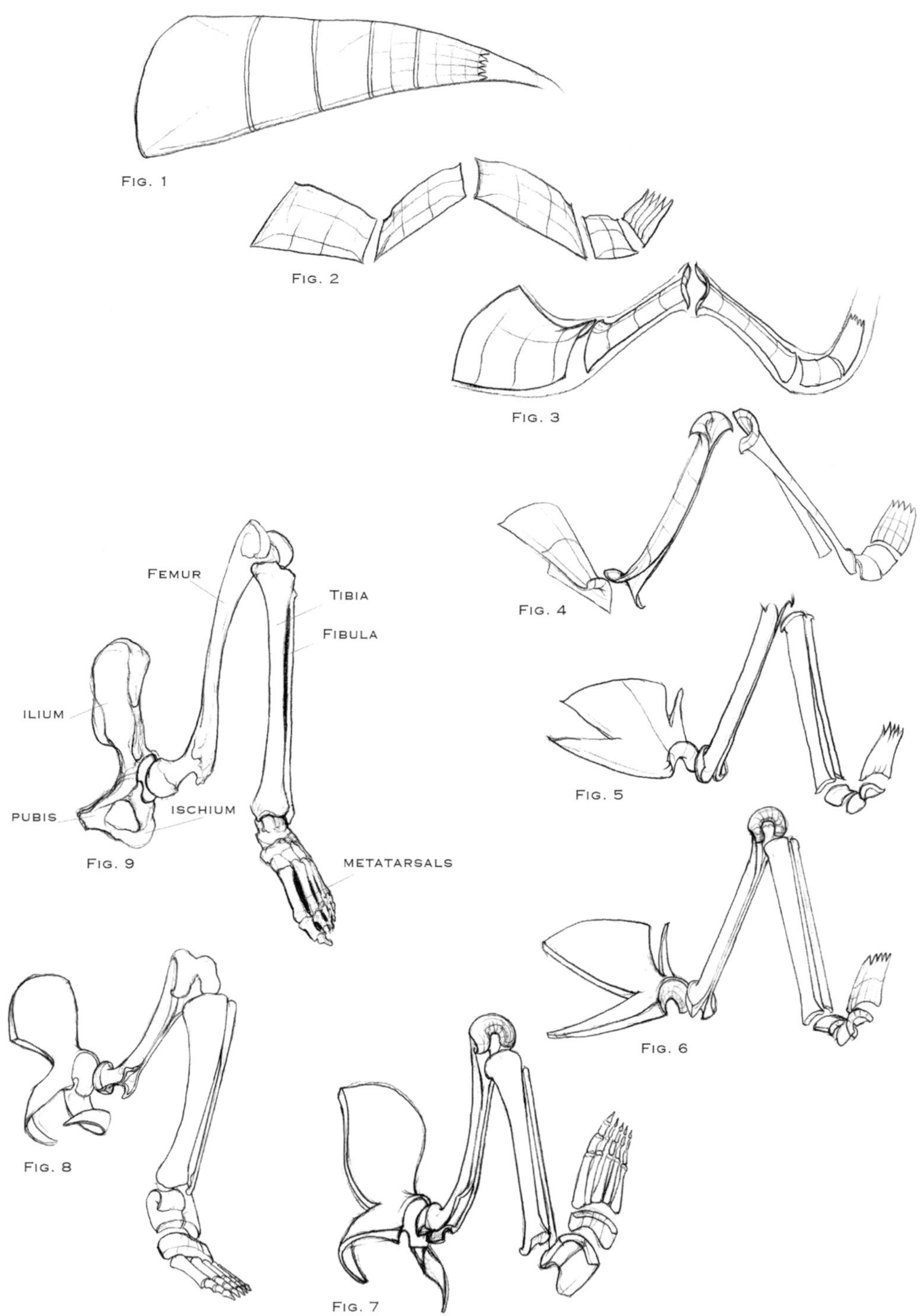

Development of the Vertebrate Limb

Plate 31

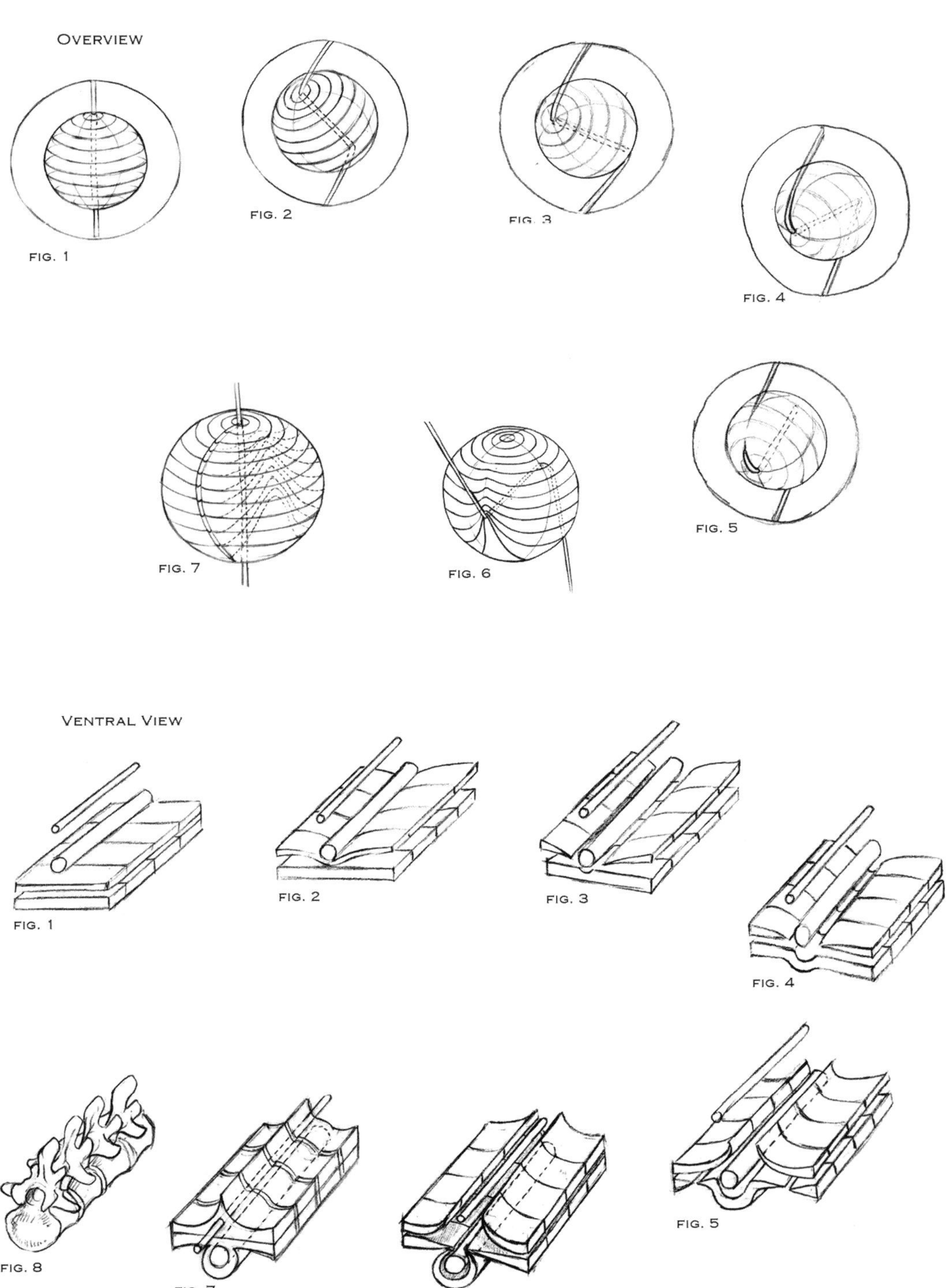

Vertebrae Formation

Plate 32

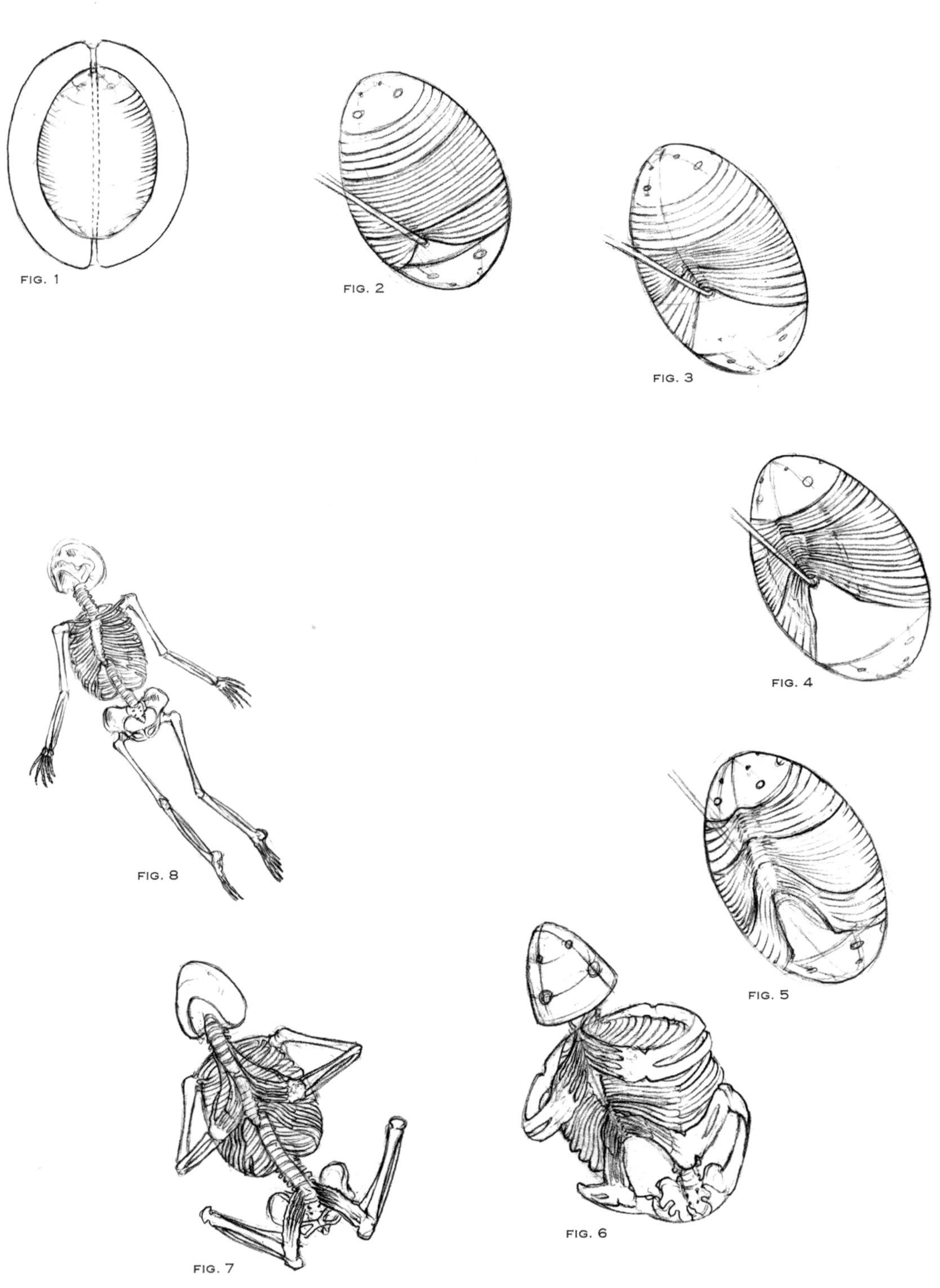

Vertebrate Embryogenesis

Plate 33

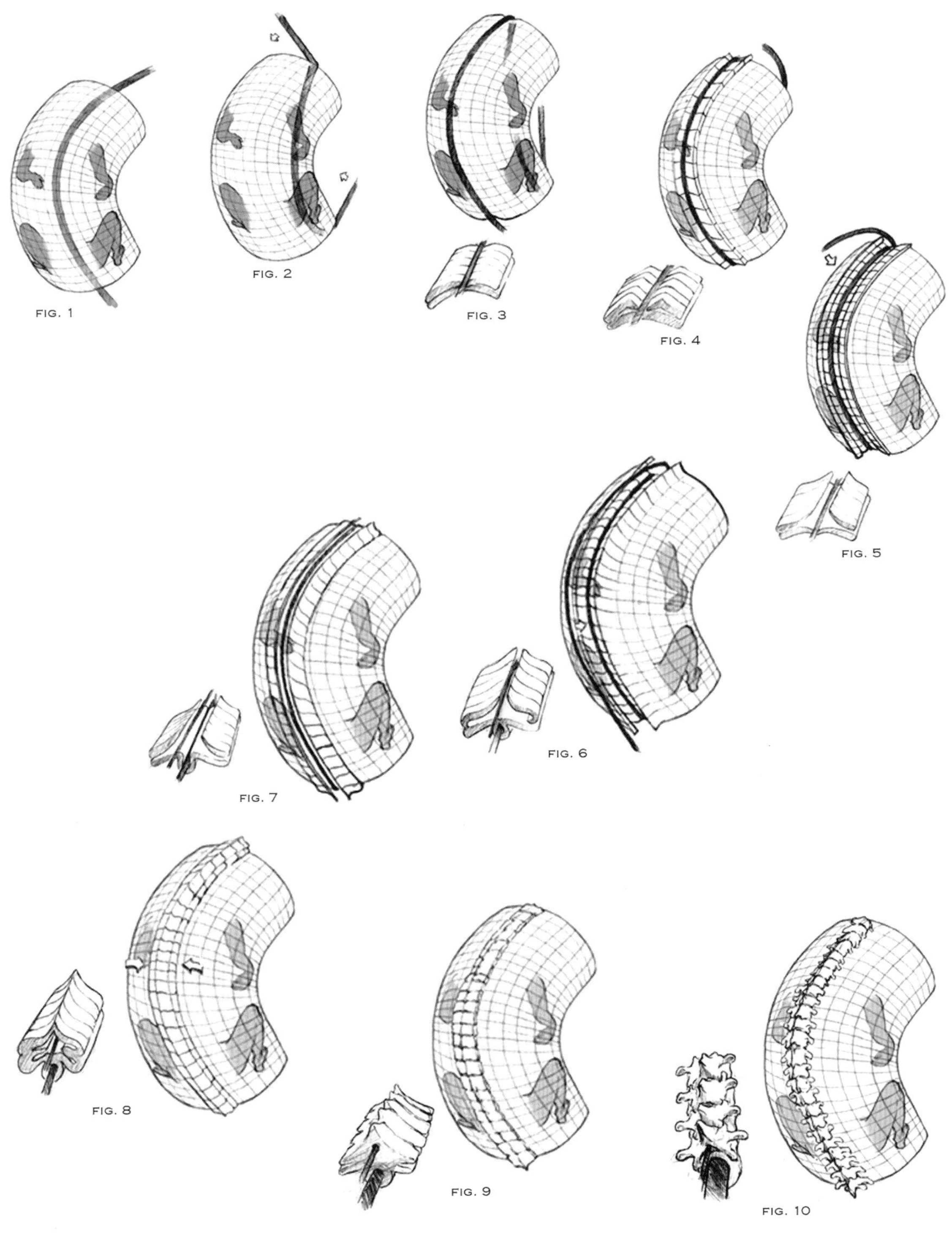

Vertebrae Formation

Plate 34

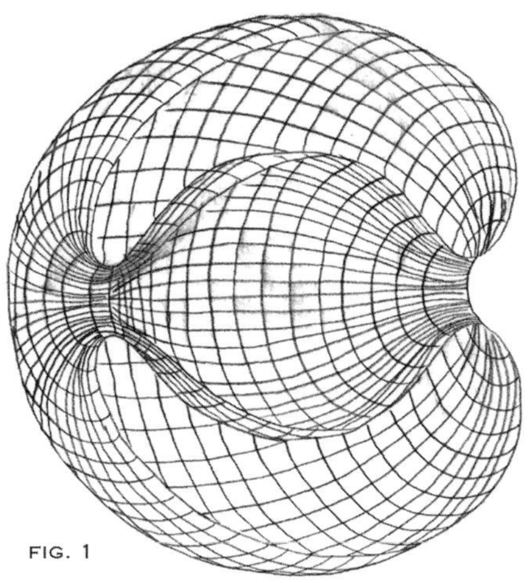

FIG. 1

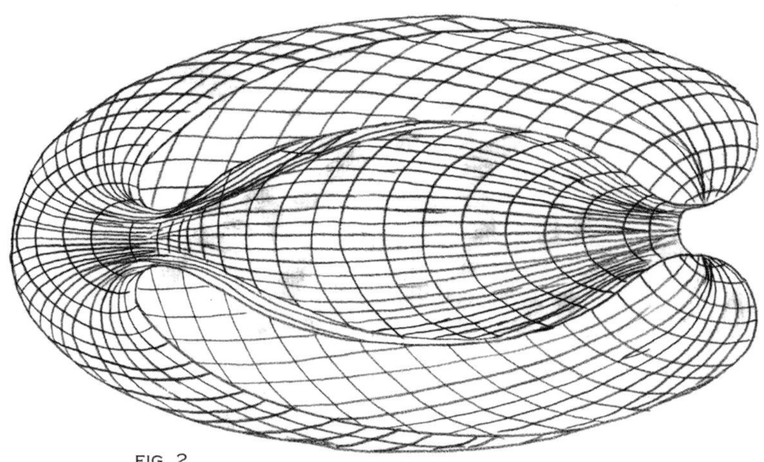

FIG. 2

THE SHAPES OF THE BONES OF THE VERTEBRATE SKELETON ARE DERIVED BY THE DEFORMATION OF A GEOMETRIC FIGURE, NAMELY, THE SPHERICAL QUADRILATERAL, OR SPHERICAL SQUARE, THE FOUR-SIDED ANALOG OF THE MORE FAMILIAR SPHERICAL TRIANGLE, PART OF THE CLASS OF POLYGONAL FIGURES INSCRIBED ON THE SURFACE OF A SPHERE, IN THE DISCIPLINE CALLED SPHERICAL GEOMETRY.

THE INSCRIPTION OF LATITUDE AND LONGITUDE LINES ON A SPHERE SUBDIVIDES THE SURFACE INTO SPHERICAL SQUARES. EACH OF THESE MAY BE DETACHED FROM THE SURFACE, CURLED AND ROLLED ALONG THE POLAR AXIS, AND THEN COMPRESSED AXIALLY. THIS MANEUVER PRODUCES THE SHAPE OF THE ARCHETYPAL VERTEBRATE LONG BONE, EASILY MIMICKED BY MODELING CLAY. THE ACTUAL OCCURRENCE OF THIS MANEUVER IN AN ANCESTRAL SPHERICAL CELL OVER EONS COULD PRODUCE A MORPHOGENETIC FIELD INHERITED BY MODERN VERTEBRATES WHICH MIMICS THIS MORPHOLOGY IN EMBRYOGENESIS, WHILE THE FIRST STAGES ARE LOST BY EVOLUTIONARY CONDENSATION.

MEMBRANE PATTERNING

PLATE 35

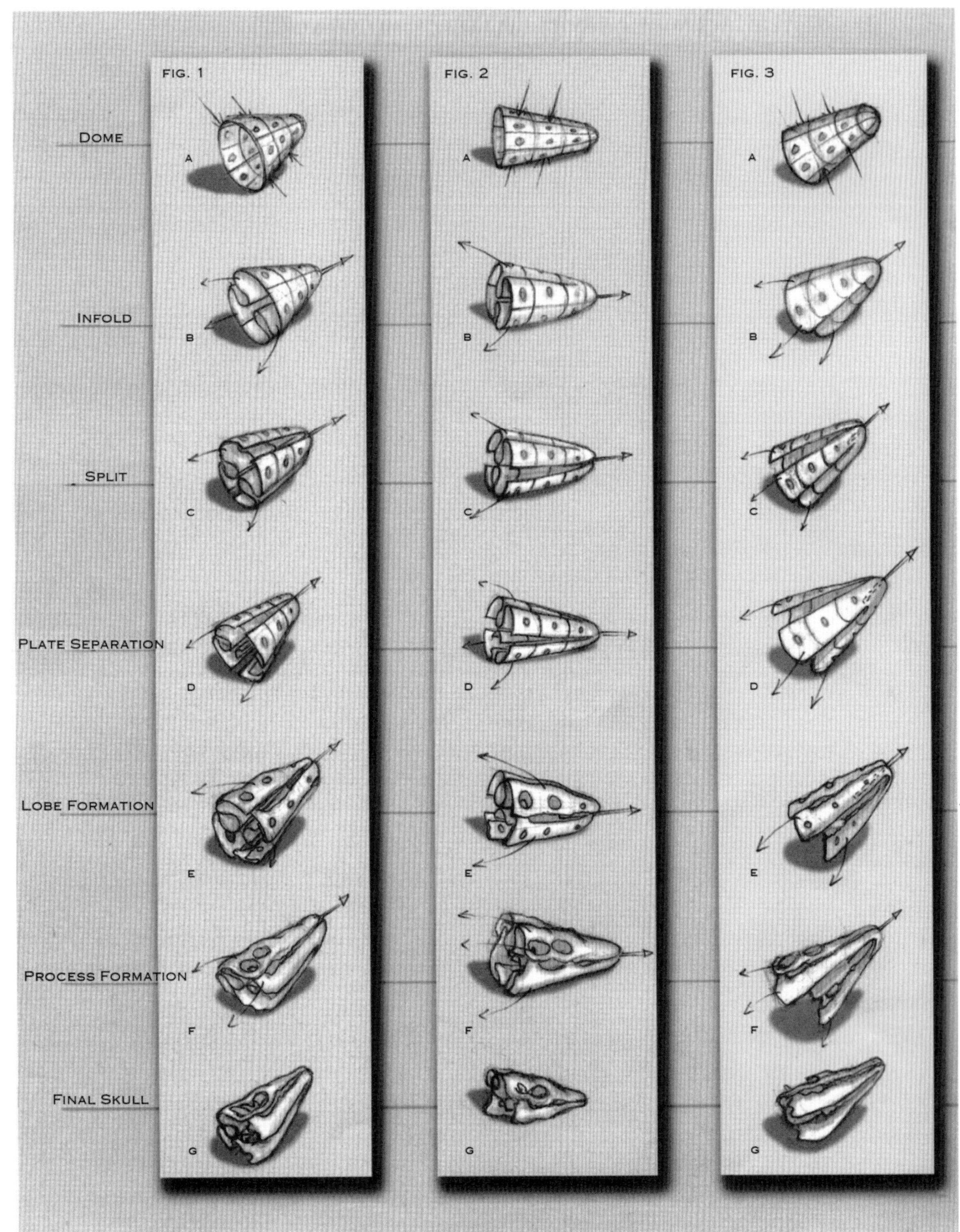

GEOMETRY OF SKULL DEVELOPMENT

PLATE 36

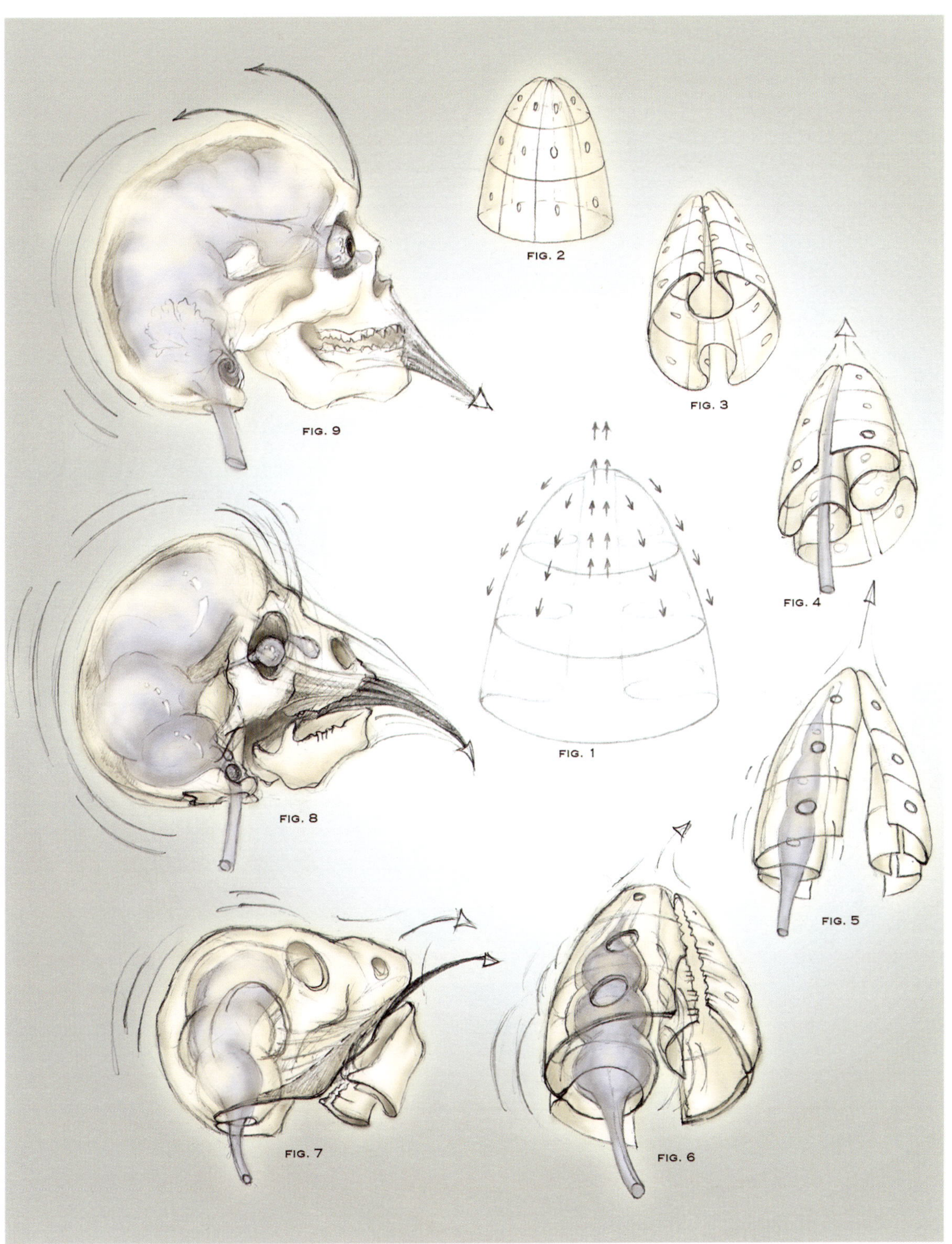

Skull Development

Plate 37

Vertebrate Gastrulation

Plate 38

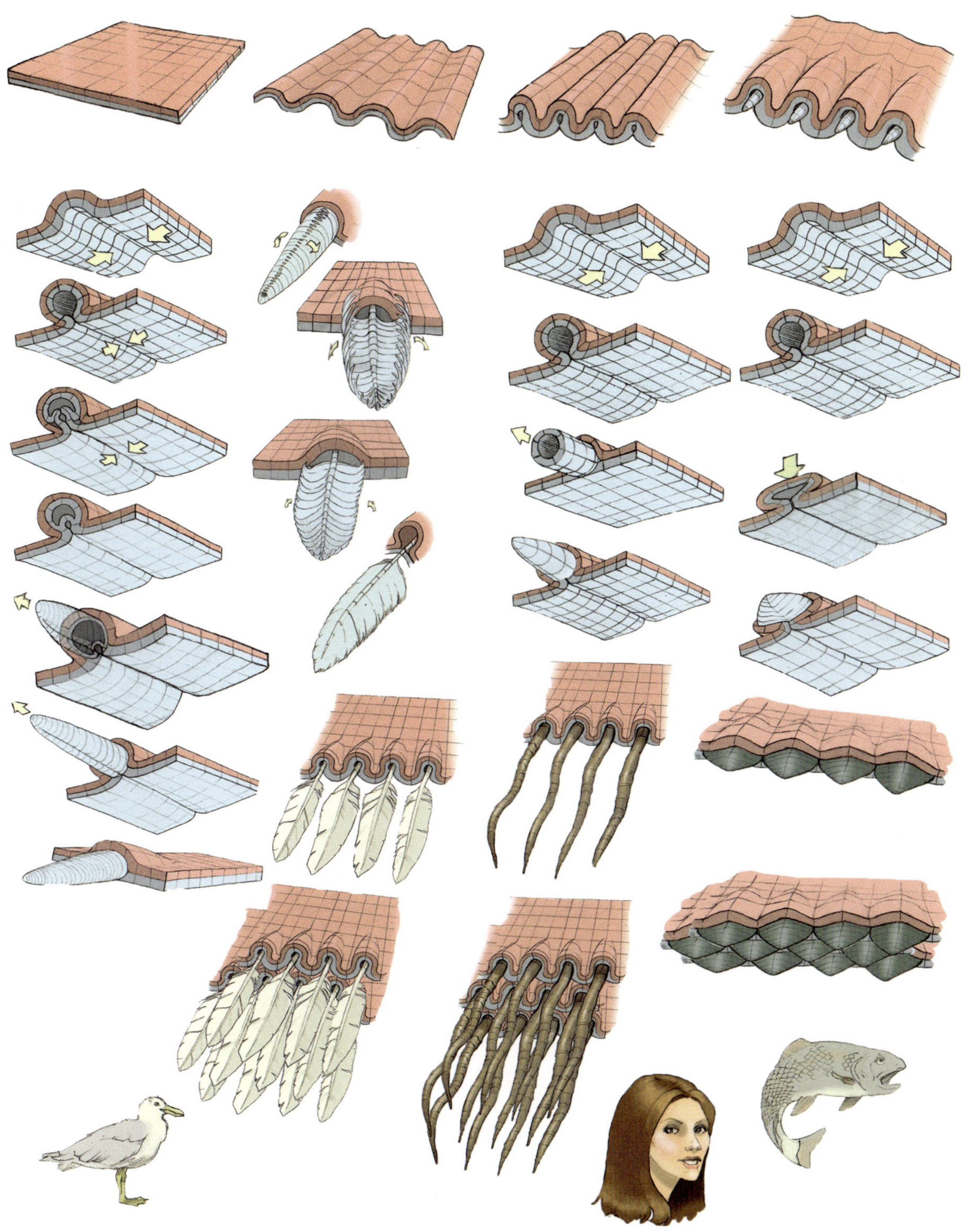

Integument Morphology

Plate 39

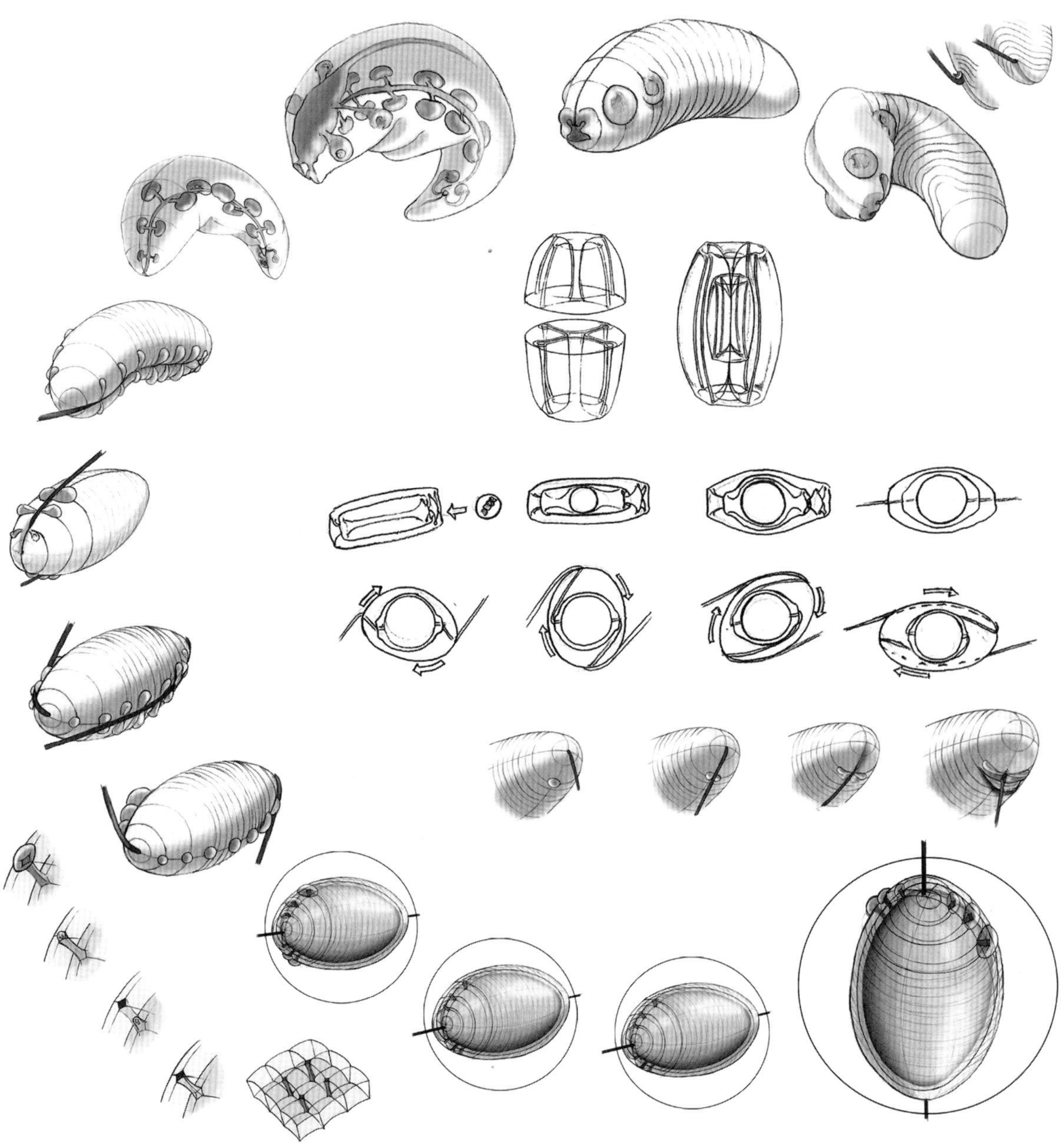

The form of the organs of bilateral animals is the observable result of the mechanical bisection of single, simple, plastic organelles occurring along the axial midlines, the separation of the identical mirror-image halves, and the concurrent growth patterns caused by mechanical deformation.

This structure is the result of the property of protoplasm to enclose itself in a bilayer membrane of phospholipid molecules which line up in geometric arrays of columns and rows. The intersections are open passages which allow the expanding, pressurized, elastic inner membrane to extrude clear through to the outside, studding the surface with geometrically regular patterns, exemplified by the forms of the simplest animals and plants.

The primordial germ plasm, in the form of a torus within a torus, ingests an ever-growing spherical body. The counter-rotation of the two causes the taut interior cable of the outside torus to carve a deep furrow into the surface of the inner torus, bisecting the self-organized structures located on its midlines.

The Human Blueprint

Plate 40A

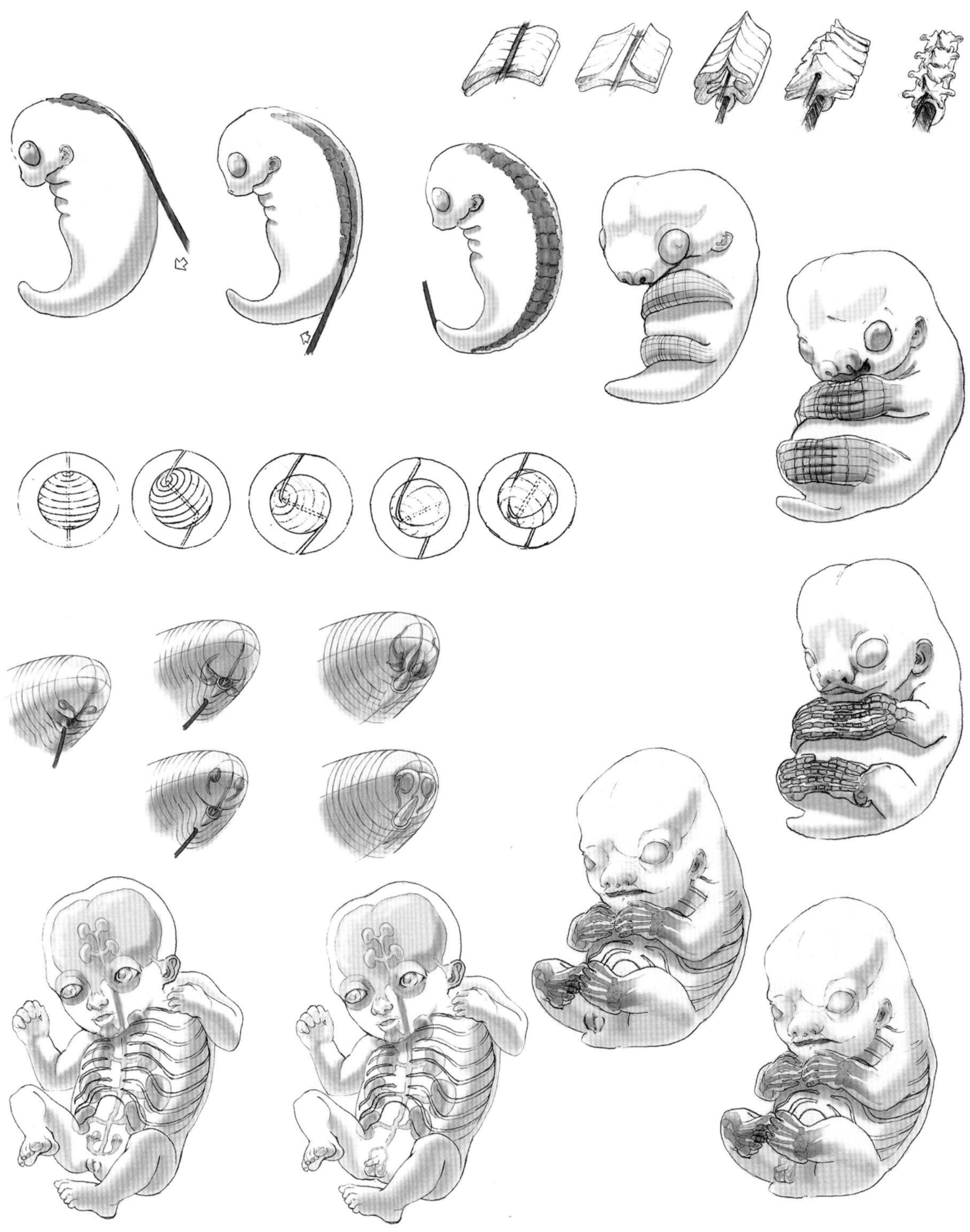

The vertebrate body plan is based on a sphere with rows of holes down opposite midlines. The taut cable of an enclosing, counter-rotating, outer toroidal membrane bisects the domed extrusions, forming the sense organs above, and the internal organs, vertebrae, and limbs below.

Plate 40b

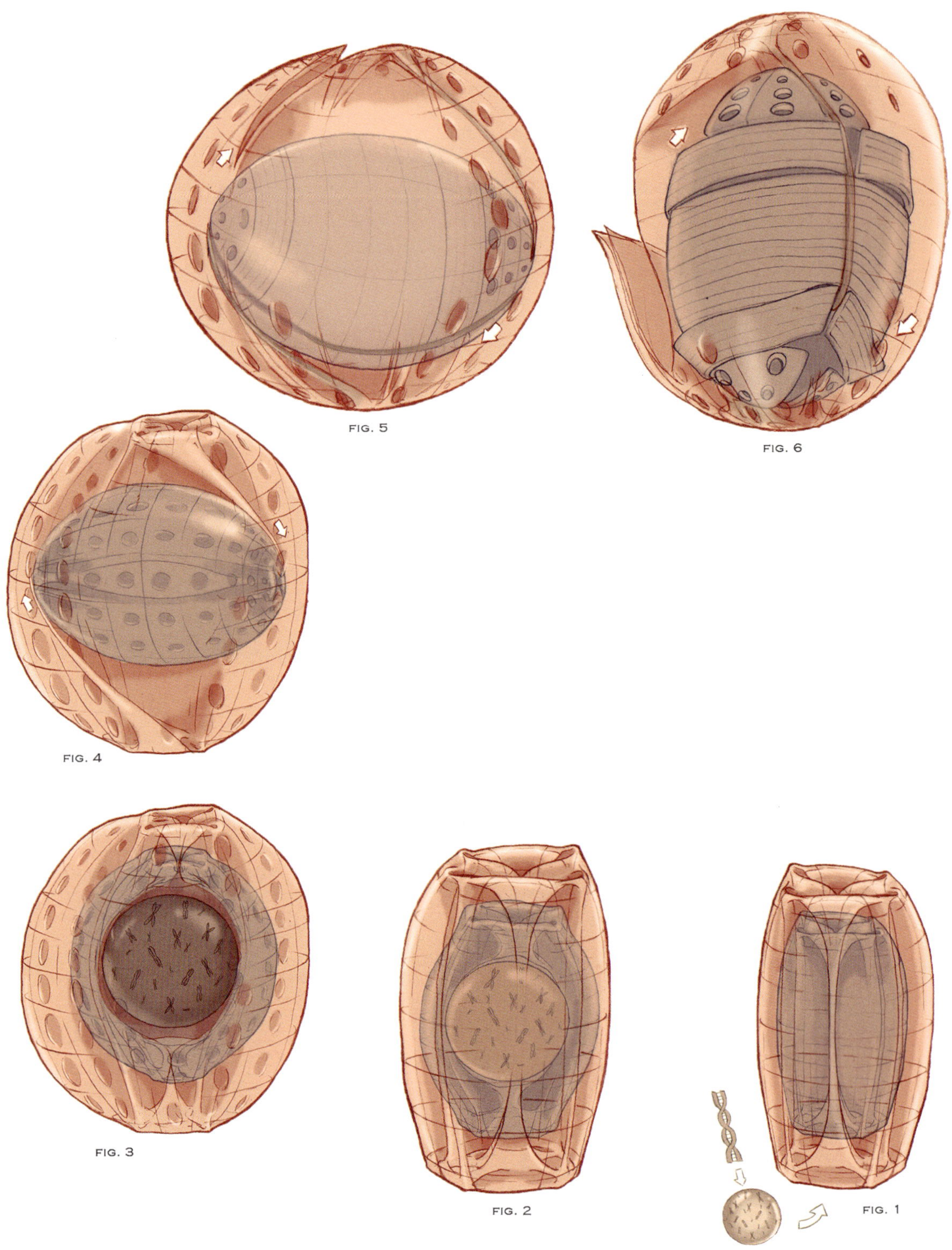

VERTEBRATE EMBRYOGENESIS

PLATE 41A

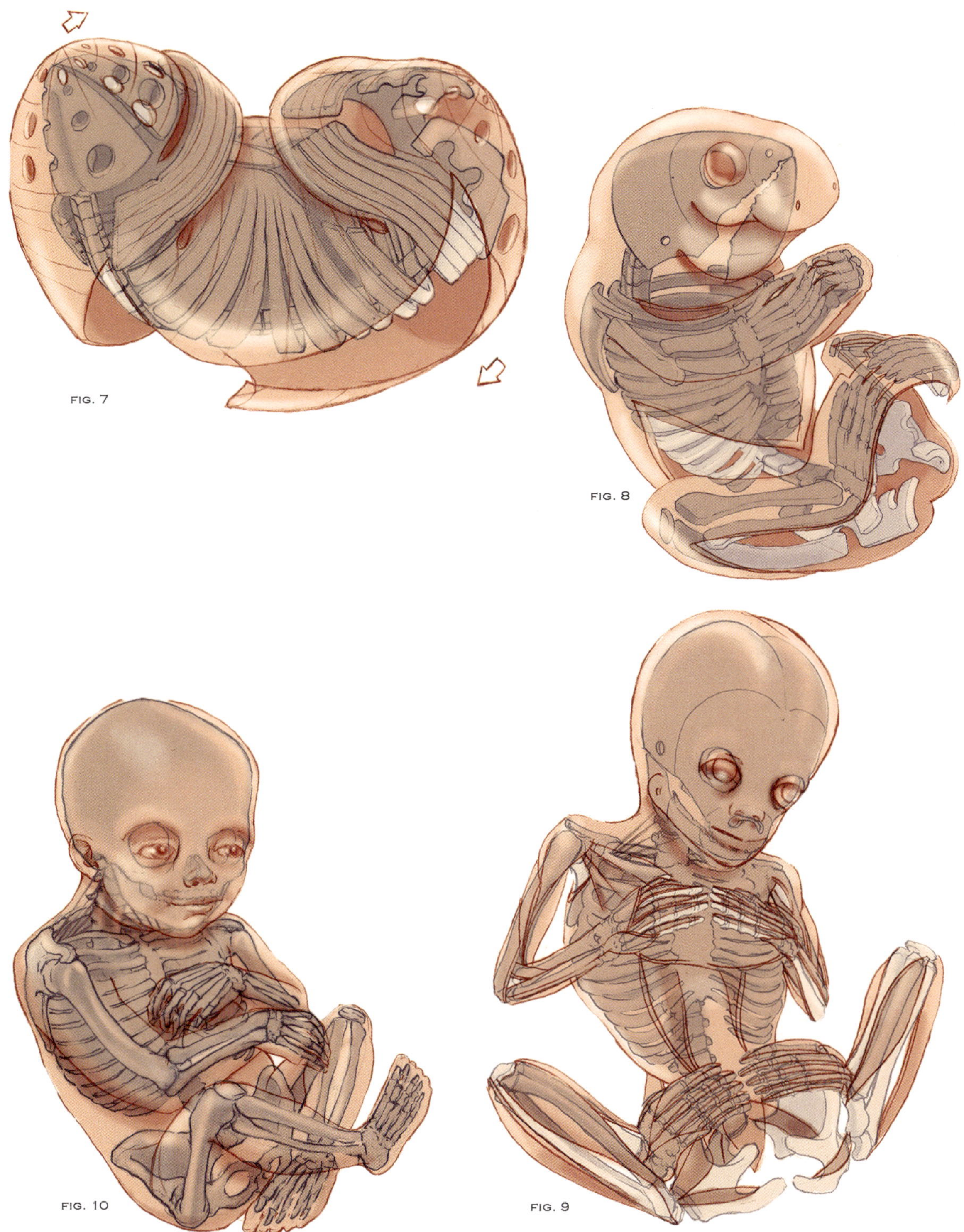

PLATE 41B

Flower Anatomy by the Deformation
of the Toroidal Form of the Germ Plasm

Plate 42a

Plate 42b

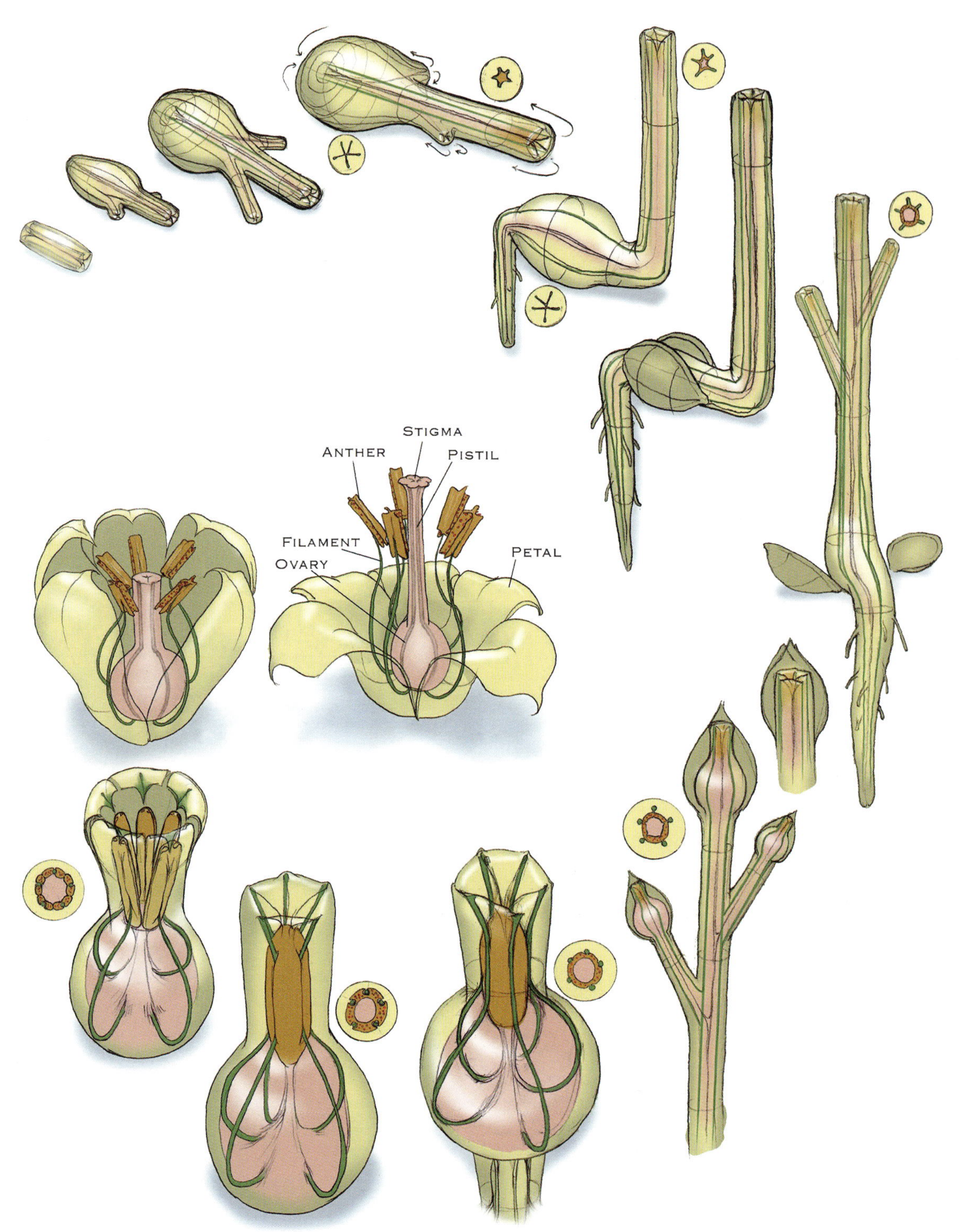

Flower Morphogenesis

Plate 43

Fruit Development

Plate 44

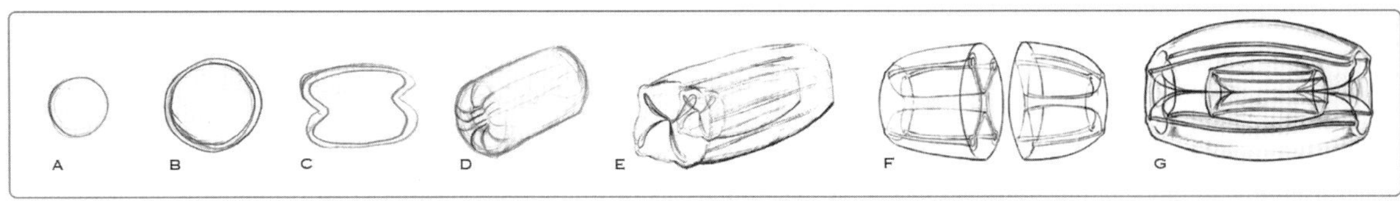

FIG. 1

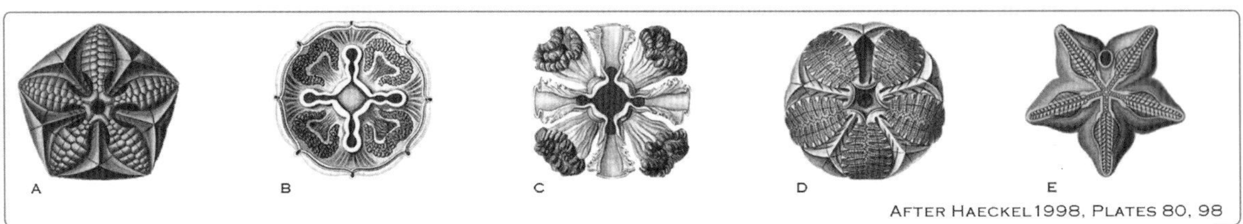

AFTER HAECKEL 1998, PLATES 80, 98

FIG. 2

FIG. 3

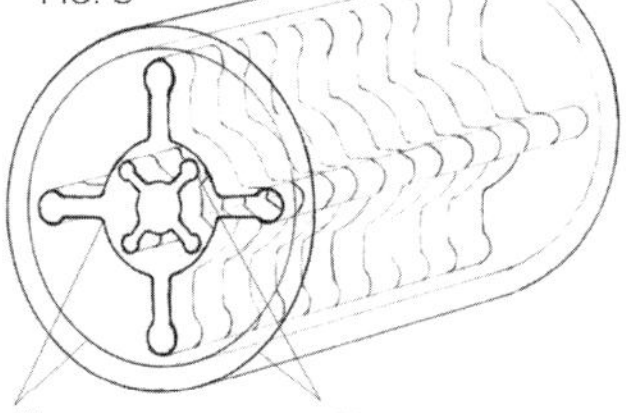

INNER TORUS

OUTER TORUS

MULTI-TORUS

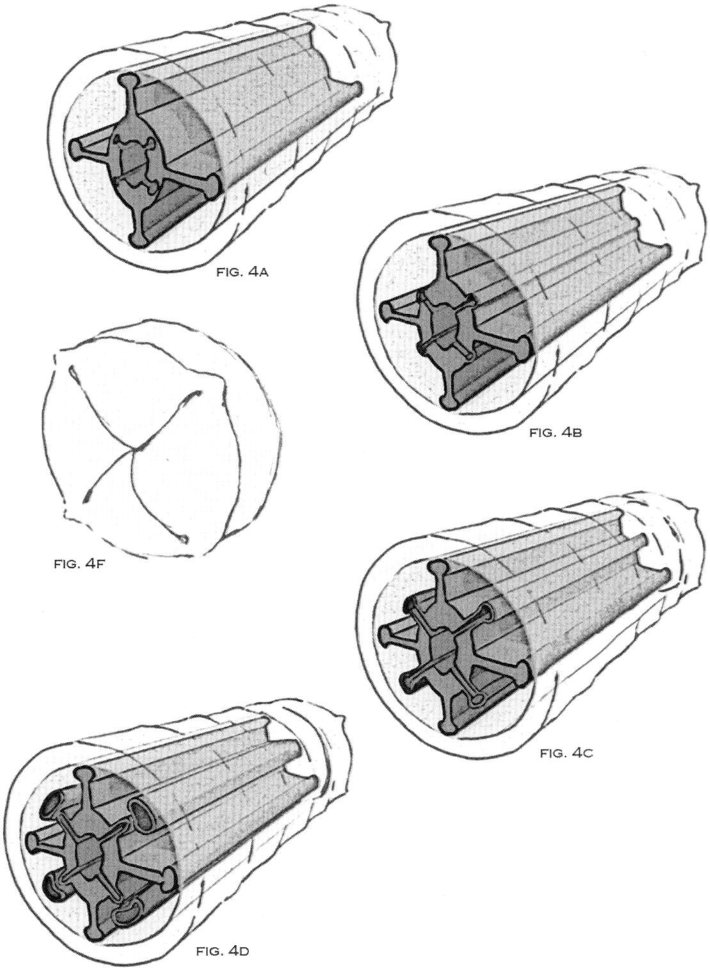

FIG. 4A

FIG. 4B

FIG. 4F

FIG. 4C

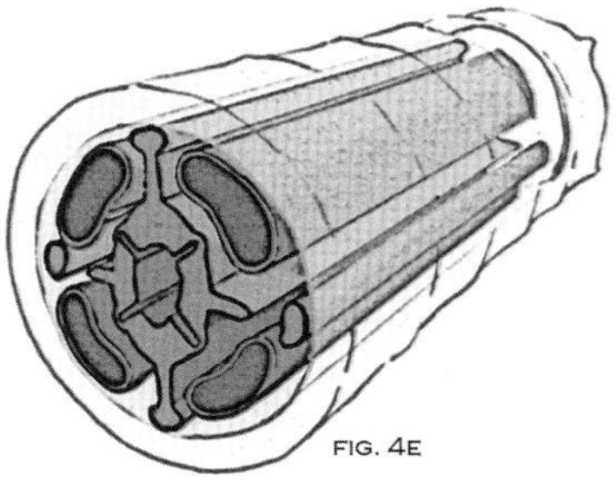

FIG. 4E

FIG. 4D

ESTABLISHMENT OF THE GERM LAYERS

PLATE 45

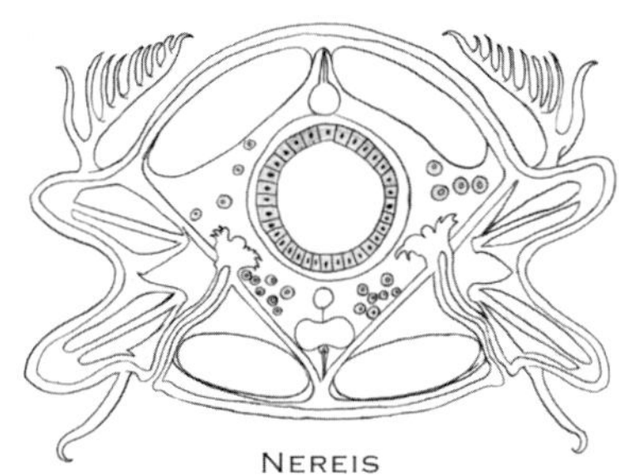

FIG. 1

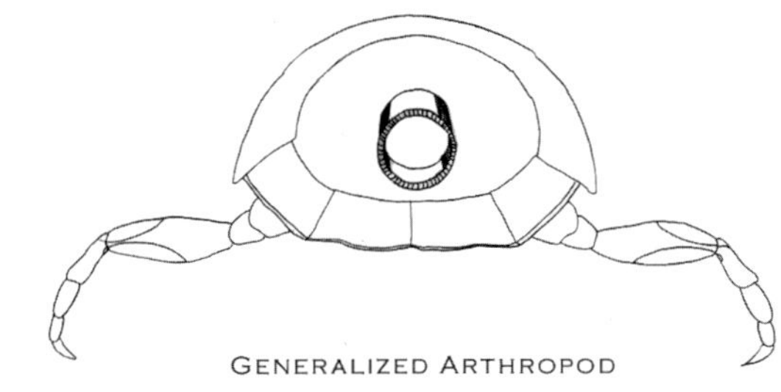

FIG. 2

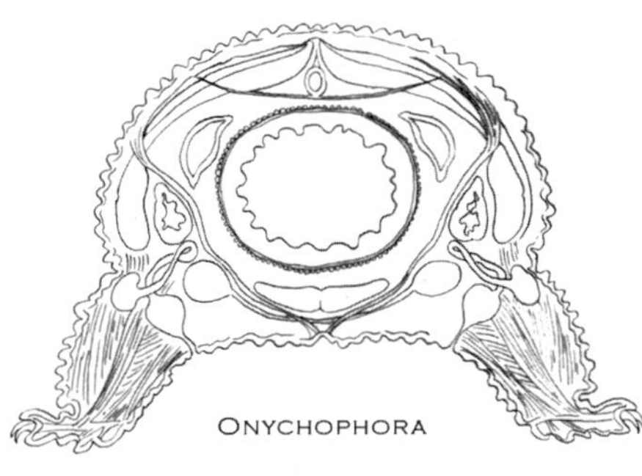

FIG. 3

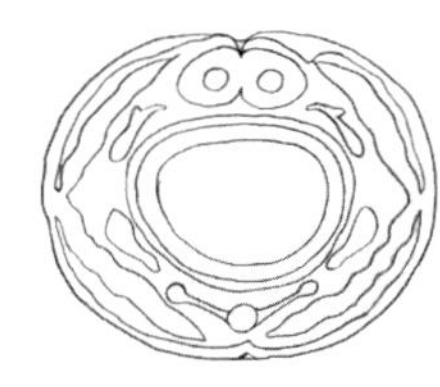

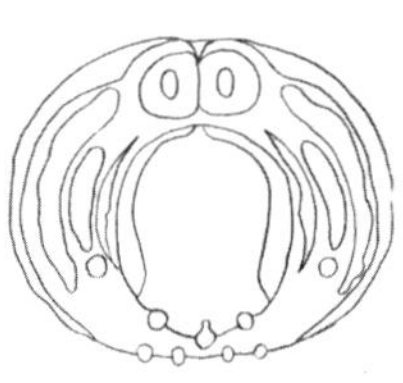

DROSOPHILA

FIG. 4

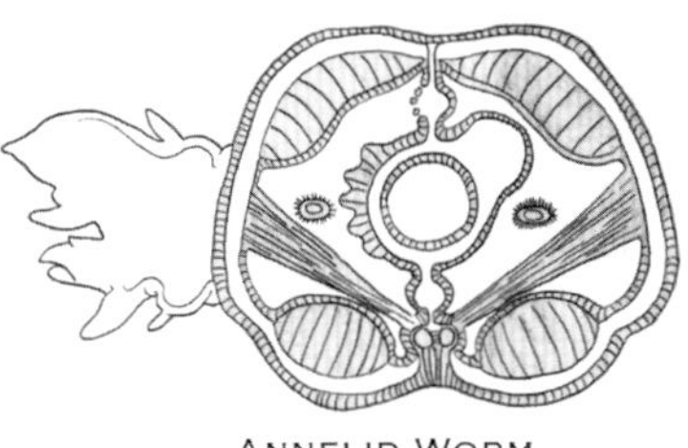

FIG. 5

FIG. 1 AFTER STORER 1943, FIG 21-12
FIG. 2 AFTER VALENTINE 2004, FIG. 7.19
FIG. 3 AFTER VALENTINE 2004, FIG. 7.12C
FIG. 4 AFTER VALENTINE 2004, FIG. 3.9B,C
FIG. 5 AFTER VALENTINE 2004, FIG. 2.14G

THE FIRST DEMONSTRATION SHOWS THE CONSEQUENCES OF THE GROWTH OF THE INNERMOST TUBE, THAT OF THE OUTER TORUS, WITH RESPECT TO THE SPACE ENCLOSING IT. THE ARMS OF THE CROSS EXTEND, DEFORMING THE CIRCULAR ENCLOSURE BY PRODUCING BULGES TO ACCOMMODATE THE EXPANSION. THE PROCESS IS SEEN TO EXTEND TO THE POINT THAT THE EXTREMITIES OF THE EXPANDING CROSS FILL THE SPACE AVAILABLE AND THEN PINCH OFF, CREATING UNATTACHED ENCLOSURES. THE RESULTING FIGURE IS A SIMULATION OF THE CROSS-SECTION OF A TYPICAL BILATERAL ANIMAL BODY FORM, SHOWING THE FORMATION OF THE THREE GERM LAYERS OF THE TRIPLOBLASTIC BODY, THE ARCHENTERON, THE ARTERIAL AND VENAL BLOOD VESSELS, THE LONGITUDINAL MUSCLES, THE LATERAL NEPHRITIC DUCTS, AND MESENTERY SEPTA WHICH DIVIDE THE TWO SIDES AND THE DORSAL FROM VENTRAL, SUBDIVIDING THE INTERIOR IN HYDRAULICALLY SEPARATE, PERITONEAL CHAMBERS.

ESTABLISHMENT OF THE GERM LAYERS

PLATE 46

Flowchart of Phyletic Differentiation by Mechanical Deformation
Plate 47

Invertebrate Phyla
Plate 48

PHYLETIC ORIGINS

PLATE 49A

Plate 49b

Flower Morphogenesis

Plate 50

Lepidopteran Wing Pattern Morphogenesis

Plate 51

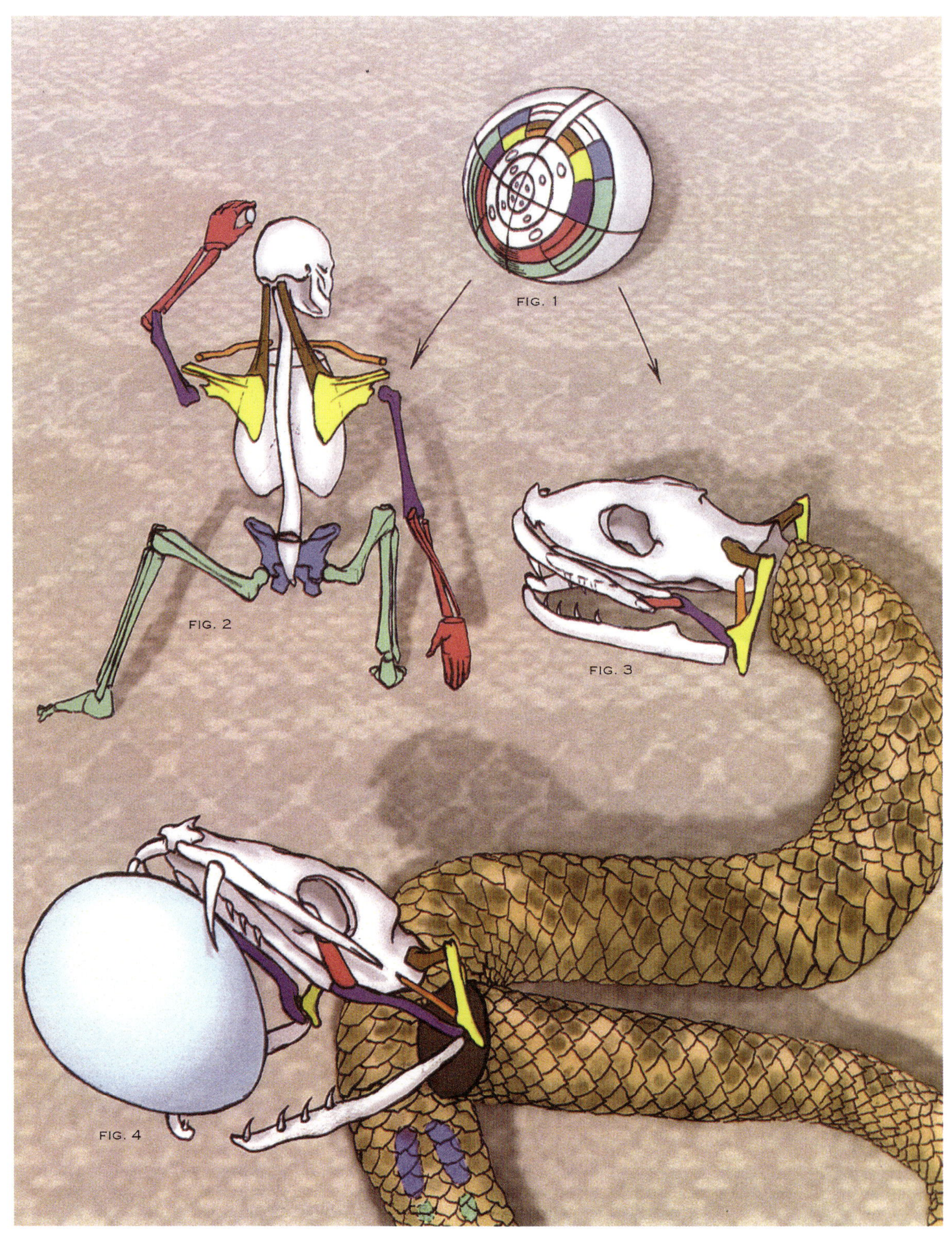

Snake Jaw Morphogenesis

Plate 52

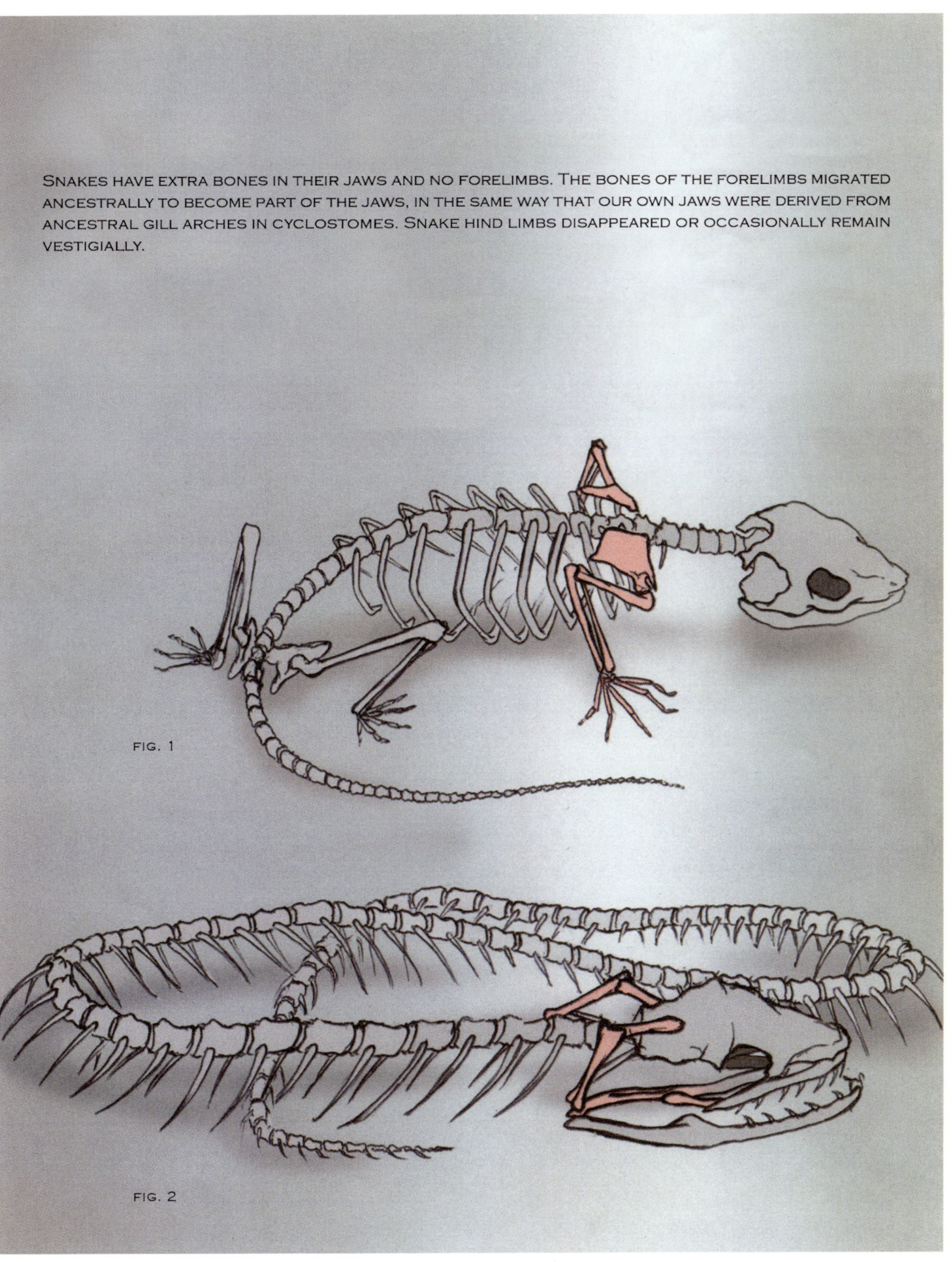

SNAKE JAW MORPHOGENESIS

PLATE 53

Geometric Origin of Turtle Shells

Plate 54

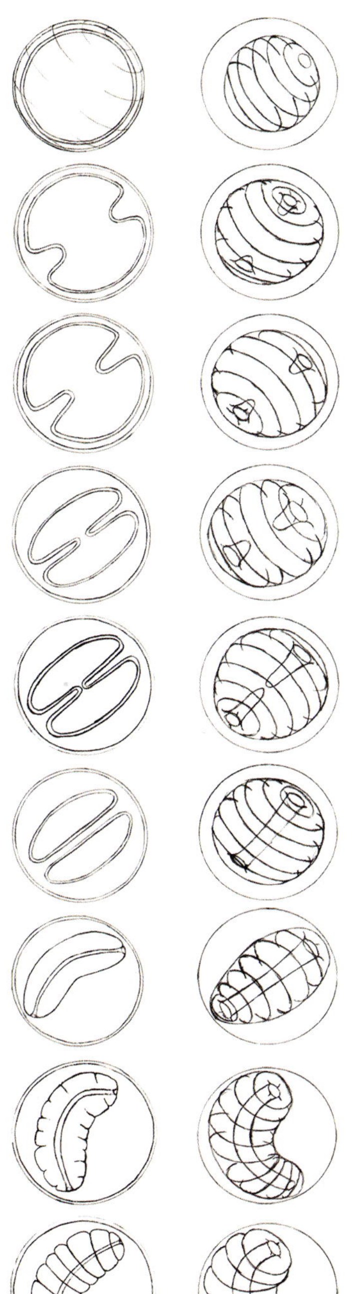

FROM SPHERE TO TORUS BY CONSTRAINED EXPANSION

In the 1920s Alexander Oparin demonstrated that cell precursors, tiny spherical bilayers, are formed automatically when you shake a solution of lipids in a test tube. In 1953 Miller and Urey synthesized life chemicals from water, methane, ammonia, carbon monoxide, and a spark plug. These two experiments are the basis of the leading theory of the origin of life. They do not, however, account for form.

It is suggested here that form begins as the continuation of the self-organization of the bilayer cell membrane. The expansion of the inner membrane of the spherical bilayer transforms it into a curled, segmented tube, the form of the archetypal embryo, from whence species development proceeds.

The initial form of the embryo of most bilateral animals, including insects and vertebrates, is the segmented worm-shape curled within a spherical shell, the familiar form of the larva of bees, beetles, flies, etc. The following is a model of how this archetypal embryo may be formed by self-organization:

> The membrane of the biological cell is a spherical bilayer consisting of a double wall of lipid molecules. New lipid molecules produced by the genes migrate to the inner membrane and join it, causing it to expand. The inner membrane expands inwardly within the constraint of the outer one. Guided by the laws of mechanics, the surface enters the interior by indentations at opposite poles. These join to form an internal canal, transforming the spherical layer into a tube, or torus. Continued expansion causes the elongated figure to shrivel in periodic rings and to curl as its length exceeds the internal diameter of the enclosing sphere. Internal creases caused by curling separate at the midline and swing outward as the figure finally straightens.

Organic form may be simulated by the expansion of a sphere within a sphere.

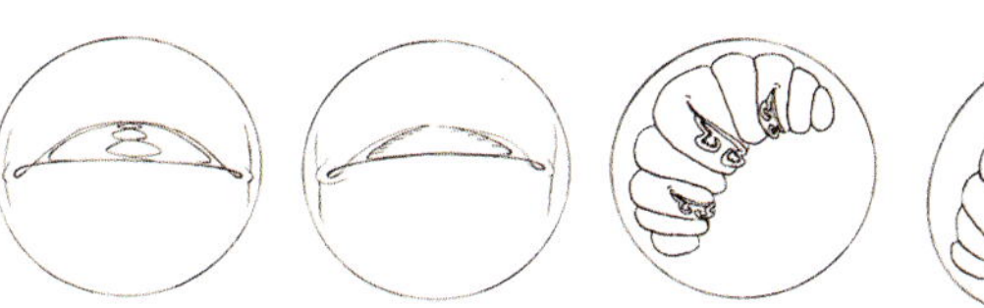

The Self-organization of the Archetypal Embryo

Plate 55

Afterword

Richard Milner

The sprouting up of "heretical" ideas on the origin of organic forms, which are the substance of many books and articles, have led me on an unexpected path. After decades of near-complacency about the evolutionary biology I thought I knew, I have found the solid ground beneath me become suddenly untrustworthy, as Darwin felt when he got too close to earthquakes in South America.

When I wrote the above paragraph in early June 2007, I had no clue that an intellectual earthquake was about to rock biology two weeks later, demolishing cherished theories about DNA and the genome, just as many scientists, among which the Harvard geneticist Richard Lewontin, had correctly predicted would eventually happen.

I was becoming increasingly skeptical of the established life sciences; even some of its trivial news releases were making me cringe. Take this newspaper clip from June 1, 2007: a biotech company had just presented James D. Watson, co-discoverer of the structure of DNA, with a transcript of his own genome sequence, which filled two CDs. He accepted them with humble graciousness, as the owner of McDonald's might react to ceremoniously receiving the umpteenth billion hamburger sold.

Promoted as the first printout of a human genome made for under $1 million, I suppose the non-event was meant to demonstrate to the public how the costs of genomic sequencing have plummeted. Was the common man expected to cheer at the imminent trickle-down of this expensive technology? Cheap genome home test kits are just around the corner. But the problem is, except for a few warning markers of disease vulnerability—many of which you could deduce from family history—would the results tell us anything useful or significant?

Writing in *Nature*, journalist Erika Chek noted that "such personal genomes (as the one made for Watson) are for now largely symbolic, because it's difficult to draw concrete information about a person's health from his or her genome." The earthquake was beginning to rumble.

Genomes are described as "largely symbolic" in *Nature,* a bastion of the biotech establishment. Were cracks in the edifice suddenly appearing in public view? Genomics dominated biology for half a century, and had become a holy grail. Many seemed to regard the genetic code, in President Bill Clinton's phrase, as "the language in which God has written the story of living things." It was sold to the public as biology's long-sought Rosetta Stone to understanding how a living thing is formed. When the preliminary completion of the genome was first announced in 2000, Francis Collins, director of the U.S. Human Genome Research Institute, stood by Clinton's side and proclaimed, "we have caught the first glimpse of our own instruction book."

The problem was: no one could read the so-called instructions, nor had they the slightest clue as to how they built a living body. Scientists and intellectuals working on theories of structural self-organization had concluded several years ago that the adoration of the genome was a scientific wrong turn that should be consigned to the dustbin of history, along with phlogiston, the Homunculus theory, spontaneous generation, and the four humours of the body.

Biology's underlying rationale, since at least the late 1950s, has been that genes and DNA are like an architect's blueprints for a house. First the code has to be spelled out in terms of its chemical components, and their sequence. Then, at some point, we could discover how this blueprint is translated into membranes, flesh, and bone. A series of Hollywood blockbuster movies was based on the premise that living dinosaurs could be reconstructed from the DNA inside a fossilized mosquito.

Alas, biologists have awakened from what Lewontin called "the dream of the human genome." True, much has been learned about DNA over the last few decades, but the function of over 95% of the genetic material was completely unknown, and called "junk DNA," a label for ignorance. Many genes seem to do nothing but regulate other genes. Some command certain formal features to be repeated many times, and others are simply on-off switches. (Do you think you can understand the form and function of the electrical appliances in your kitchen by studying their on and off switches?) Other genes speed up development or slow it down. Some jump around and change places in the sequence, and others reinterpret old patterns in new ways. One very odd thing is that more complex creatures do not necessarily have more genes than simpler ones; a sea urchin for instance, has almost as many genes (23,000) as a person does.

All that accumulation of facts is worthwhile, but many believe it has distracted biologists from the main prize: the origin of form. By the turn of the twenty-first century, however, it was becoming clear (though few in the biotech enterprise were prepared to admit it) that the emperor wasn't wearing any clothes. Biologists were stymied about the great unspoken question; how do these genes and the proteins they produce become a plant or animal? Some maverick biologists in several countries broke with the prevailing paradigm to pursue work related to a biomechanical view of life, an alternative explanation of how, in Charles Darwin's phrase, "the birds and the beasts are formed."

If the genome is indeed a blueprint, then how come no one has achieved even a glimpse of how flesh and blood organisms are built from the genomic plans? It would be as if, on the analogy of an architect's blueprints, no one ever saw the foreman reading the plans, materials being ordered and gathered, and the workmen at their tasks. No one has yet seen the genomic hod carriers, or brick layers, or roofers, or builders at work. Back in 1888, the German embryologist Wilhelm His accepted August Weismann's new ideas about the "germ plasm" with this caveat: "I should be the last to discard the law of organic heredity … but the single word 'heredity' cannot dispense science from the duty of making every possible inquiry into the *mechanism* of organic formation. To think that heredity will build organic beings without mechanical means is a piece of unscientific mysticism."

On June 14, 2007, the earthquake finally hit. The venerable AFP news agency announced that, after four years of study by 35 scientific groups from around the world (joined together for the ENCODE project), the scientists were forced to conclude that their model of the genome was wrong, and that "a cornerstone concept about the chemical code for life is badly flawed."

Scrutinizing a small percentage of the genome, the scientists had tried to identify the role of every component in forming an organism. They could not even begin to accomplish their self-appointed task and concluded "that an established theory about the genome should be consigned to history."

In the old view, in between the genes (which comprise only a twentieth of the genetic code) are vast stretches of so-called "junk" DNA, which were thought to be inert evolutionary leftovers. But ENCODE (Encyclopedia of DNA elements) concluded that the genome is actually "a highly complex, interwoven machine with very few inactive stretches," and the mislabeled "junk" has a crucial role in manufacturing proteins. As science journalist Richard Ingham lyrically reported to the AFP news agency, "Previously written off as silent, it emerges as a singer with its own discreet voice, part of a vast, interacting molecular choir."

Most of the genome now appears to be transcribed into RNA, which relays information from the DNA to the cellular machinery. That is a remarkable finding, since it had been thought that only a fraction of the genome was transcribed. Also, it seems to be composed of elements that have no discernable benefits for survival and reproduction, and shows no signs of having been sculpted and winnowed by natural selection.

Back in 2000, Lewontin published *It Ain't Necessarily So: The Dream of the Human Genome and Other Illusions.* Lewontin said, among other things, that the accepted genomic model was "wrong in what it claims to explain. First, DNA is not self-reproducing; second, it makes nothing; and third, organisms are not determined by it." Lewontin asserted that DNA is "a dead molecule, among the most nonreactive chemically inert molecules in the living world," is produced by a complex cellular machinery of proteins, and does not itself produce proteins.

It's hard not to imagine that the scientists at the Human Genome Project had conducted a massive preemptive strike against potential critics. Four years ago, when they realized that their theoretical model would not stand for long, they formed a group within the scientific establishment to knock it down themselves and emerge as self-correcting scientists rather than wasters of billions of dollars. Whenever that happens, scientists can always save face. They were not to be seen as a pack of blinded, self-deluded hyenas so much as humble participants in "the self-correcting nature of science."

What will rise out of this rubble? What new theories of how organic form is produced will compete to be the new reigning paradigm? To understand the work done by proponents of structural self-organization regarding how embryos and forms originate, we need to revisit one of the heroes of the movement, the nineteenth-century embryologist Wilhelm His.

After studying embryos of various animals in the quest for the origins of form, His found that in their early stages they behaved like bubbles and elastic tubes and bladders. Adopting an experimental approach, he created models in rubber and other flexible materials, and bent, twisted, and inflated them. Sure enough, some of them mimicked the folding and topologies of early-stage embryos. During his lifetime, His was bullied by Ernst Haeckel, who denigrated his methods as *Gummischlauchwissenschaft* (rubber bladder science). But now, a century and a half later, there are many who think he was on the right track, and they have picked up where the Goethian embryologists had left off in their quest for the bio-mechanical principles and organic origami that lie behind the forms of all living things. Indeed, scientists who base their work on the idea of self-organization—that matter contains within itself the ability to organize into spontaneous patterns, like crystals or snowflakes, without the help of genomes or Darwin—are growing in number.

The Goethian quest for the universal factor that generates all organic form is rooted in a long tradition of rational morphology, which had its roots in the eighteenth-century century Enlightenment, and strongly influenced such philosophers and thinkers as Kant,

Goethe, St. Hilaire, Cuvier, Owen, Bateson, and D'Arcy Thompson. The view of spontaneous transformation and origins of form that has been adopted by many intellectuals has absolutely nothing to do with mystical creationism, and everything to do with algorithms, fluid mechanics, topology and other such concepts. Darwinism triumphed, as Lewontin and others have pointed out, not because it was the complete explanation of evolution and the origin of species, but because it attempted to do so within a mechanistic, naturalistic framework—a continuation of the tradition of Laplace, Newton, and Descartes.

One of the pioneers in structural self-organization is Stuart Kauffman, a theoretical biologist and astrophysicist at the University of Calgary, who has written that "Self-organization is a natural property of complex genetic systems. There is 'order for free' out there, a spontaneous crystallization of order out of complex systems, with no need for natural selection or any other external force." Kauffman further opines that "If the new science of complexity succeeds, it will broker a marriage between self-organization and selection. It will be a physics of biology."

As a graduate student in anthropology at the University of California (Los Angeles and Berkeley) forty years ago, I was raised in the Neo-Darwinian Synthetic Theory, promulgated in the 1950s by Ernst Mayr, Julian Huxley, Theodosius Dobzhansky, and others. These brilliant men sought to integrate many disparate studies, including animal behavior, paleontology, comparative anatomy, and population genetics all under the same tent of evolution by natural selection. Their model of evolution was based on the notion that 1) organisms in populations continually produce a wide range of variations, including many random gene mutations, 2) these can vary in almost infinite directions, with equal randomness, 3) natural selection creates form by winnowing these genes in a non-random (adaptive) manner, and 4) that this takes place gradually over immense periods of time, in a long succession of slightly intermediate forms.

There are two problems with this formulation: 1) Natural selection was repeated endlessly as a mantra, even in cases where it could not be demonstrated to have occurred. Indeed, experimental demonstrations of natural selection from the 1950s through the end of the twentieth century were few and far between. We were all taught about the peppered moths in England that turned from white to black in the polluted woods, and Bumpus' sparrows, whose median wing shape and lengths were favored in surviving storms, but (apart from a few bacteria and fruitflies) no one saw a new species evolve before their eyes. 2) The second problem was that their view of timing and variation proved to be erroneous. Organisms do NOT vary randomly and all over the place. Variations tend to be constrained and biased—the same ones coming up again and again, and some possibilities NEVER come up. Even in "monstrosities" (what Cuvier named "teratologies") like three-eyed sheep and six-legged pigs, the same deformities reappear from time to time, the result of similar biased mistakes in development. Moreover, the fossil record seems to indicate that there are long periods of very little change (stasis), followed or (in Stephen Jay Gould's phrase) "punctuated" by fairly rapid spurts of evolution. Animals still evolve, and long periods of time are still involved, but the kind of gradualism Darwin usually assumed has not been supported by a century of genetics and paleontology.

Oh, no! Natural selection not the be-all and end-all of evolution? In fact, Darwin knew quite well that natural selection does not create form. It winnows out forms that are less successful, but is not the engine that generates such basic patterns as bilateral or radial symmetry, as in the contentious events of the Cambrian "explosion," when all the forms of modern animals first appeared. Citing natural selection as the "cause" of evolution does not at all address the question of how forms arise, what proportions are useful, and how they develop both in individuals and in species. Natural selection may choose, shape, and direct, I have come to see, but it cannot create. That was my smaller, personal earthquake.

Some years ago, a couple of evolutionary biologists made a study of how toy makers had arrived at the cute face of the most popular teddy bears. Perceptions of cuteness are tied to the facial proportions of human infants—large eyes, very high forehead, small nose and jaws. Toy makers found that the more they enlarged the eyes and shrunk the snouts of their bears, the more they would sell. Eventually, the faces of teddy bears (and Disney and Japanese animated characters) came to resemble human infants. The buyers were not creating or designing the forms, but they were nevertheless shaping their features by consistently selecting them.

Darwin's champion and "bulldog" Thomas Henry Huxley, who crusaded tirelessly for the ideas of evolution and descent with modification, never felt completely at ease with natural selection. And sad to say, many biologists continue to uphold it as the main driving force of evolution out of habit, and would rather not raise doubts for fear that religious creationists will claim that scientists know "Darwin is wrong." In the creationist's limited two-choice universe, if "Darwinism" is an incomplete theory, the only alternative to account for the origins of form is biblical literalism.

Speaking of juvenile facial features, the importance of infantilization appears to be of greater significance than I ever supposed. It is known as neoteny, a slowing down of the rate of development that makes adult humans retain the proportions of juvenile apes. We no longer sport the heavy brow ridges, large canine teeth, and projecting snouts of gorillas or chimps; instead, there has been a change in the timing of human development.

It has been shown that Steve Gould, whom I thought was an arch proponent of natural selection, harbored grave doubts about its universal efficacy. In his brilliant (but slyly subversive) book *Ontogeny and Phylogeny*, Gould unequivocally stated that "Neoteny has been a (probably *the*) major determinant of human evolution.... Human development has slowed down ... adaptive features of ancestral juveniles are easily retained." What surprised me even more was to discover that the importance of neoteny was also trumpeted and championed at length over the years by a distinguished group of scientists, including Karl von Baer, Gavin de Beer, Walter Garstang, J. B. S. Haldane, Julian Huxley, George Gaylord Simpson, and Ashley Montagu.

Blind adherence to authority in science, as Albert Einstein said, is bad science. Thomas Henry Huxley told his students that he would rather they question authority and explore new views than to parrot his lectures back to him. Indeed, Huxley said, "When science adopts a creed, it commits suicide." It's hard to accept constant change and uncertainty about the truth—but that is the soul and mandate of science, and also its excitement—the never-ending quest. But when you've got a lab and research funding at stake, it's not easy to question the basis of what keeps you in business.

R. Milner
Associate in Anthropology
American Museum of Natural History

NEOTENY AND HETEROCHRONY

STUART PIVAR

"Neoteny has been a (probably *the*) major determinant of human evolution...."
Stephen Jay Gould, *Ontogeny and Phylogeny*

That one single, simple, phenomenon called neoteny, the retardation of embryological development, can be the cause of the entire gamut of characteristics which set off humans from our ape ancestors is surely the most compelling effect in human evolution. At once it accounts for our upright posture, large brain, free-thinking curiosity, hairlessness, flat face, flat feet, appositive coital position, and dozens of other uniquely human characters. Almost as amazing is the fact that, although the effect was identified a hundred years ago, and published by the foremost biologists since, most recently by Gould and Montagu, it is ignored by the biological establishment in its summary dismissal of all things inconsistent with the dogma of the Modern Synthesis and remains unfamiliar to most students of anthropology.

In his landmark work, *Ontogeny and Phylogeny* (1977), Gould discusses the mechanistic tradition of embryology and evolution, clearly expressing his belief in evolution by endocrinal change in the process called heterochrony. Following is a very brief view of this complex subject.

Mutations consist of a change in the proportions of the existing, unchanging, generic body form by a change in the rate of growth or development of the parts. The shape of the embryo is the result of a race between growth and development. In ontogeny each part of the organism develops and grows at a rate independent of the rest. The genes maintain constancy. Random errors cause mutants which rarely survive. If development is retarded and growth remains the same then the adult form is juvenilized. This effect is called neoteny or paedomorphosis. Retarded growth and accelerated development produce a more developed adult form.

The Cambrian phyla were completed 450 million years ago. Subsequent evolution is the result of varying the proportions of the organs by alternating neoteny and acceleration. No novel forms ever appear. Since the generic phyla were fully developed, the evolutionary escape route is in the direction of neoteny, as indicated by the sequence of Picaia, placoderm, fish, amphibian, crocodilian reptile, dinosaur, mammal, primate, hominid, H. sapiens. The ratio of the size of the head to the rest of the body increases in evolution and decreases in development. In evolution the straight spine of the adult becomes curved while in embryology the reverse happens. These phenomena confirm phylogenic neoteny. The robustness of species, resulting from the utter perfection of the reproduction of the parent form, unchanged over millions of years, is the result of the eugenic effect of natural selection. Every organ in the embryology of an individual is subject to the retardation or acceleration of development, producing an oversized, undersized, or absent organ. These are almost always lethal at an early stage.

In the darkness of the cave, natural selection fails to cull the animal with sightless eyes. The neotenous absence of melanin produces the albino mutation which in darkness does not lead to vulnerability like the peppered moth. Cave animals are usually blind and albino. Bats, as well as being blind, have in many species contorted facial morphology congruent with embryonic stages. Moles have the same facial anomaly. Blind cave fish invariably have presumptive eyes in the form of an early stage of normal development. The skull is distorted as space left vacant is occupied by surrounding tissue, including the olfactory primordia, resulting, by coincidence, in a useful, heightened sense of smell, which will persist, protected against mutant drift by natural selection. Natural selection may maintain, but not originate, form.

The Direction of Vertebrate Evolution

The Hypothesis that Post-Cambrian Evolution Is the Juvenilization of the Ready-made Cambrian Phyla

It has for long been observed that the human is a juvenilized ape, the result of neoteny, the retention of juvenile, or embryonic, features in the adult. Likewise, it may be noted that apes are juvenilized mammals, mammals are juvenilized reptiles, which are juvenilized fish, which are juvenilized, acephalic pre-chordates, dating back to the Cambrian vertebrate ancestor Picaia.

The history of the vertebrate phylum is one of persistent juvenilization of an initially fully developed adult Cambrian animal. The evidence is in these morphological trends:

- Over evolutionary time the spine begins straight and becomes arched. A thin elongated body becomes rotund, a small head becomes large, the dorsal face becomes ventral, horizontally disposed limbs rotate to the vertical.

- In development each of these trends are reversed. The embryo starts with a large head and the embryonic curl and ends up with a small head and straight spine, dorsal face (except humans), and vertical legs. Neoteny preserves embryonic features in the adult, which is why these traits are reversed evo and devo, or ontogeny and phylogeny. This trend is seen in insects and crustaceans as well. Compare the primitive apterygotic silverfish with a beetle.
- The persistent, net effect of neoteny accounts for this overarching evolutionary trend. The dinosaurs and birds are a neotenous version of the early crocodilian-type reptiles and amphibians. In birds an oversized yolk presses and flattens the forelimbs dorsally and the hind limbs ventrally. Seals and walruses reverse this trend, where boneless hind limbs are flattened like wings. Amphibian limbs are under-developed fins.

Retardation and Neoteny in Human Evolution

The Seeds of Neoteny

Stephen Jay Gould

With the consummate arrogance that only an American millionaire could display, Jo Stoyte set out to purchase his immortality. Dr. Obispo, his hired scientist, discovered that the fifth earl of Gonister had, by daily ingestion of carp guts, prolonged his life into its third century. Stoyte and Obispo rushed to England, broke into the earl's quarters and discovered to Stoyte's horror and Obispo's profound amusement that man in his allotted three score years and ten is but an axolotl in its pond. We are neotenic apes and the fifth earl had grown up:

> "A foetal ape that's had time to grow up," Dr. Obispo managed at last to say. "It's *too* good!" Laughter overtook him again. "Just look at his face!" he gasped ... Mr. Stoyte seized him by the shoulder and violently shook him ... "What's happened to them?" "Just time," said Dr. Obispo airily. Dr. Obispo went on talking. Slowing up of developmental rates ... one of the mechanisms of evolution ... the older an anthropoid, the stupider ... the foetal anthropoid was able to come to maturity ... It was the finest joke he had ever known. Without moving from where he was sitting, the Fifth Earl urinated on the floor.

So wrote Aldous Huxley in his novel *After Many a Summer Dies the Swan*—a few years after his brother Julian's important work on delayed metamorphosis in amphibians and on the heels of Louis Bolk's fetalization theory of human origins.

If a good public press is the sign of a theory's impact, then the proponents of human neoteny should be satisfied with the suffusion of their theory into popular consciousness—though its influence can only be described as modest relative to that once wielded by the opposing concept of recapitulation (Chapter 5). At least, human neoteny has motivated a long defense of Rudolf Steiner's mysticism (Poppelbaum, 1960) and played an important role in several treatises in the current fad for "pop ethology" (Jonas and Klein, 1970; Shepard, 1973).

Excerpted from *Ontogeny and Phylogeny,* The Belknap Press, 1977, pp. 352–358.

The notion of human neoteny has its roots in two obvious facts: the striking resemblances between juvenile pongids and adult humans and the obliteration of this similarity during pongid ontogeny by strong negative allometry of the brain and positive allometry of the jaws (Fig. 61). Louis Bolk did not discover these phenomena in his fetalization theory of the 1920s; they had been recognized as soon as juvenile pongids had reached the zoos and museums of Europe. Moreover, the subject was not merely treated as an incidental observation; for Étienne Geoffroy Saint-Hilaire (1836a, 1836b), in describing the form and behavior of a young orang-utan recently brought to Paris, placed his observations firmly in the context of recapitulatory theory.

Geoffroy began by noting the extent of ontogenetic allometry and retracting his earlier conclusion, based on museum skeletons, that young and old orangs represented two separate genera:

> Could we have dared hope in 1798 that such different crania, one taken from a juvenile, the other from an adult, would reveal the fact of successive development in a single species? For we have a distance greater than that between the genera *Canis* and *Ursus.* (1836a, p. 94)

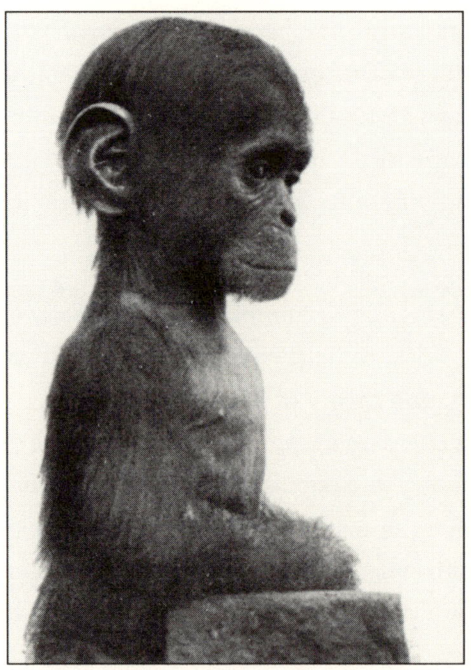

Fig. 61. Baby and adult chimpanzee, after Naef, 1926b. Naef remarks: "Of all animal pictures known to me, this is the most manlike" (p. 448).

He then attributes these allometries to the early cessation of growth in the orang brain:

> The skull of a young orang strongly resembles that of a human child. The cranial vault, which faithfully represents the form of an organ it protects, could be taken for that of a small human were it not for the more forward position of the maxillaries and the much larger cutting teeth. But it happens, as a result of the progress of age, that the contents practically cease to increase while the container grows strongly and continually. (1836b, pp. 6–7)

Geoffroy extended his observations to the habits and behavior of his young orang:

> In the head of the young orang, we find the childlike and gracious features of man … We find the same correspondence of habits, the same gentleness and sympathetic affection, also some traits of sulkiness and rebellion in response to contradiction … on the contrary, if we consider the skull of the adult, we find truly frightening features of a revolting bestiality [*formes vraiment effroyable et d'une bestialité révoltante*]. (1836a, pp. 94–95)

Yet humans display an "inverse system of development" (1836b, p. 7), for our brain does not so quickly cease its growth: both form and behavior remain close to the juvenile state. Could human ontogeny represent an "arrest of development" with respect to the next lower link in the chain of being? Geoffroy preferred to save recapitulation by regarding the adult orang as anomalous—as a form developed "too far" in its own ontogeny and therefore not recapitulated during human development: "This new revelation of such a great departure from the rules [of recapitulation] that we have discovered is a teratological fact of the greatest importance" (1836a, p. 94).

The resemblance of adult humans to juvenile apes was treated as an anomaly throughout the heyday of recapitulation. But single ugly facts, despite Huxley's aphorism, do not destroy great theories. Recapitulationists simply followed Haeckel's strategy in arguing that the anomalies were few, recognizable, and unimportant (Chapter 6). But man posed a special problem: if the most "important" of all species had evolved by retardation, then exceptions to recapitulation could not all be relegated to insignificance. E. D. Cope considered the problem in great detail and admitted that many human features had evolved by retardation. But he quickly added that these retarded features were not involved in our superiority, and that the progressive features of our mental development displayed acceleration and recapitulation:

> I have pointed out that in the structure of his extremities and dentition, he agrees with the type of Mammalia prevalent during the Eocene period. Hence in these respects he resembles the immature stages of those mammals which have undergone special modifications of limbs and extremities … I have also shown that in the shape of his head man resembles the embryos of all Vertebrata, in the protuberant forehead, and vertical face and jaws. In this part of the structure most Vertebrata have grown further from the embryonic type than has man, so that the human face may be truly said to be the result of retardation. Nevertheless, in the structure of his nervous, circulatory, and for the most part, of his reproductive system, man stands at the summit of the Vertebrata. It is in those parts of his structure that are necessary to supremacy by force of body only, that man is retarded and embryonic. (1896, pp. 204–205)

Cope's defense of the mainstream prevailed, and recapitulatory thinking dominated concepts of human evolution until the collapse of Haeckel's doctrine itself (Chapter 6). In the best known of more recent usages, Wood Jones sought man's origin among the tarsioids, arguing that the early ontogenetic fusion of the maxillary-premaxillary suture precluded an origin from monkeys, apes, and even from australopithecines (for these forms retain the suture longer). Writing about the fusion of the two bones, Wood Jones deemed it "probable that its early ontogenetic accomplishment is a guarantee of its early phylogenetic acquirement" (1929, p. 319).

The Fetalization Theory of Louis Bolk

The collapse of Haeckel's biogenetic law and the rise of Garstang's neoteny virtually guaranteed that a century of observations would be gathered to yield a paedomorphic theory of human origins. J. Kollmann (1905), inventor of the term "neoteny," paved the way with a curious theory that humans had originated from pygmies who had simply retained their juvenile features during phyletic size increase. The pygmy progenitors, Kollmann added, probably arose from juvenile apes that had lost the ancestral tendency to regress (*zurücksinken*) during ontogeny to lower levels of cephalization: "The juvenile orang-utan is doubtlessly better qualified for human ancestry than the ape of Trinil" (1905, p. 19).[1]

[1] Thus, Dubois's Pithecanthropus was debarred from human ancestry as a disharmonious type with perfected bipedal locomotion but too small a brain. This insistence on correlated modification of the entire body within true phyletic lineages also underlay Bolk's concept of fetalization and, from our perspective, produced most of its problems.

Louis Bolk (186–930), professor of human anatomy at Amsterdam developed the inevitable idea in a long series of papers (1915, 1923, 1924, 1926a, 1926c, 1929, for example), culminating in his pamphlet of 1926, *Das Problem der Menschwerdung*. Bolk's theory of fetalization set the stage for all later discussion. The subsequent debate has been murky and confused because Bolk's arguments have been presented outside the context of his philosophical positions. His insight has been ridiculed in the light of modern doctrine and dismissed in toto because he linked valid and important data to evolutionary views now rejected. In presenting this detailed exposition of Bolk's views, I am trying to establish a ground for the acceptance of his basic notion by referring the old observations that lie at its core (and that date at least to Geoffroy) to a philosophical context of current orthodoxy. I shall try to support three statements: (1) It is irrelevant that Bolk's evolutionary theory seems outdated or even foolish today; his theory was reasonable in his time and he supported it cogently. (2) The data that he presented can survive the collapse of his explanatory structure. (3) We must try to identify the "philosophical baggage" that underlies all theories—both to understand why a man says what he does and to aid in rescue operations when new philosophies require a separation of baby from bath water.

I want to rescue Bolk's data—and his basic insight—from the evolutionary theory to which he tied it and from which it has not been adequately extracted.

Bolk's Data

To support the argument that we evolved by retaining juvenile features of our ancestors, Bolk provided lists of similarities between adult humans and juvenile apes: "Our essential somatic properties, i.e. those which distinguish the human body form from that of the other Primates, have all one feature in common, viz they are fetal conditions that have become permanent. What is a transitional stage in the ontogenesis of other Primates has become a terminal stage in man" (1926a, P. 468). In his most extensive work Bolk (1926c, p. 6) provided an abbreviated list in the following order:

1. Our "flat faced" orthognathy (a phenomenon of complex cause related both to facial reduction and to the retention of juvenile flexure, reflected, for example, in the failure of the sphenoethmoidal angle to open out during ontogeny).
2. Reduction or lack of body hair.
3. Loss of pigmentation in skin, eyes, and hair (Bolk argues that black peoples are born with relatively light skin, while ancestral primates are as dark at birth as ever).

4. The form of the external ear.
5. The epicanthic (or Mongolian) eyefold.
6. The central position of the foramen magnum (it migrates backward during the ontogeny of primates).
7. High relative brain weight.
8. Persistence of the cranial sutures to an advanced age.
9. The labia majora of women.
10. The structure of the hand and foot.
11. The form of the pelvis.[2]
12. The ventrally directed position of the sexual canal in women.
13. Certain variations of the tooth row and cranial sutures.

To this basic list, Bolk added many additional features; other compendia are presented by Montagu (1962), de Beer (1948, 1958), and Keith (1949). The following items follow Montagu's order (pp. 326–327) with some deletions and additions:

14. Absence of brow ridges.
15. Absence of cranial crests.
16. Thinness of skull bones.
17. Position of orbits under cranial cavity.
18. Brachycephaly.
19. Small teeth.
20. Late eruption of teeth.
21. No rotation of the big toe.
22. Prolonged period of infantile dependency.
23. Prolonged period of growth.
24. Long life span.
25. Large body size (related by Bolk, 1926c, p. 39, to retardation of ossification and retention of fetal growth rates).[3]

---

[2] These last two points are, I confess, vague and confusing. But I have translated Bolk's words literally.

[3] This is a fairly complete list of the "basic" features usually presented to support the contention that humans are paedomorphic. Other characters are usually the concomitants or the causes of features presented above (much, for example, has been written about the role of various cranial flexures in setting the position of the foramen magnum and the disposition of the face and jaws.)

These lists from Bolk and Montagu display the extreme variation in type and importance of the basic data presented by leading supporters of human neoteny. There are obvious difficulties. The features, for example, are not all independent: the position of the orbits (17) is not unrelated to the expansion of the brain (7); prolonged infantile dependency (22) and prolonged growth (23) are aspects of a more basic retardation in temporal development. Moreover, two very different phenomena are mixed together: the slowing down of developmental rates (8, 20, 22, 23, 24) and the retention of juvenile shapes (all others).

The Biology of Belief

Bruce H. Lipton, PhD

In his 1859 book, *The Origin of Species*, Darwin said that individual traits are passed from parents to their children. He suggested that "hereditary factors" passed from parent to child *control* the characteristics of an individual's life. That bit of insight set scientists off on a frenzied attempt to dissect life down to its molecular nuts and bolts, for within the structure of the cell was to be found the heredity mechanism that controlled life.

The search came to a remarkable end fifty years ago when James Watson and Francis Crick described the structure and function of the DNA double helix, the material of which genes are made. Scientists finally figured out the nature of the "hereditary factors" that Darwin had written about in the nineteenth century. The tabloids heralded the brave new world of genetic engineering with its promise of designer babies and magic bullet medical treatments. I vividly remember the large block print headlines that filled the front page on that memorable day in 1953: "Secret of Life Discovered."

Like the tabloids, biologists jumped on the gene bandwagon. The mechanism by which DNA controls biological life became the Central Dogma of molecular biology, painstakingly spelled out in textbooks. In the long-running debate over nature vs. nurture, the pendulum swung decidedly to nature. At first DNA was thought to be responsible only for our physical characteristics, but then we started believing that our genes control our emotions and behaviors as well. So if you are born with a defective happiness gene, you can expect to have an unhappy life. (pp. xx–xxi)

What activates genes? The answer was elegantly spelled out in 1990 in a paper entitled *Metaphors and the Role of Genes and Development* by H. F. Nijhout. (Nijhout 1990) Nijhout presents evidence that the notion that genes control biology has been so frequently repeated for such a long period of time that scientists have forgotten it is a hypothesis, not a truth. In reality, the idea that genes control biology is a supposition, which has

---

Excerpted from *The Biology of Belief,* Hay House, 2008.

never been proven and, in fact, has been undermined by the latest scientific research. Genetic control, argues Nijhout, has become a metaphor in our society. We want to believe that genetic engineers are the new medical magicians who can cure diseases and while they're at it create more Einsteins and Mozarts as well. But metaphor does not equate with scientific truth. Nijhout summarizes the truth: "When a gene product is needed, a signal from its environment, not an emergent property of the gene itself, activates expression of that gene." In other words, when it comes to genetic control, "It's the environment, stupid." (pp. 21–22)

Geneticists experienced a comparable shock when, contrary to their expectations of over 120,000 genes, they found that the entire human genome consists of approximately 25,000 genes. (Pennisi 2003a and 2003b; Pearson 2003; Goodman 2003) More than eighty percent of the presumed and *required* DNA does not exist! The missing genes are proving to be more troublesome than the missing eighteen minutes of the Nixon tapes. The one-gene, one-protein concept was a fundamental tenet of genetic determinism. Now that the Human Genome Project has toppled the one-gene for one-protein concept, our current theories of how life works have to be scrapped. No longer is it possible to believe that genetic engineers can, with relative ease, fix all our biological dilemmas. There are simply not enough genes to account for the complexity of human life or of human disease.

I may sound like Chicken Little shouting that the genetics sky is falling. However, you don't have to take my word for it. Chicken Big is saying the same thing. In a commentary on the surprising results of the Human Genome Project, David Baltimore, one of the world's preeminent geneticists and a Nobel Prize winner, addressed the issue of human complexity (Baltimore 2001):

"But unless the human genome contains a lot of genes that are opaque to our computers, it is clear that we do not gain our undoubted complexity over worms and plants by using more genes.

"Understanding what does give us our complexity—our enormous behavioral repertoire, ability to produce conscious action, remarkable physical coordination, precisely tuned alterations in response to external variations of the environments, learning, memory, need I go on?—remains a challenge for the future."

As Baltimore states, the results of the Human Genome Project force us to consider other ideas about how life is controlled. "Understanding what does give us our complexity ... remains a challenge for the future." The sky *is* falling.

In addition, the results of the Human Genome Project are forcing us to reconsider our genetic relationship with other organisms in the biosphere. We can no longer use genes to explain why humans are at the top of the evolutionary ladder. It turns out there is not much difference in the total number of genes found in humans and those found in primitive organisms. Let's take a look at three of the most studied animal models in genetic research, a microscopic nematode roundworm known as *Caenorhabditis elegans*, the fruit fly, and the laboratory mouse.

The primitive *Caenorhabditis* worm serves as a perfect model for studying the role of genes in development and behavior. This rapidly growing and reproducing organism has a precisely patterned body comprised of exactly 969 cells and a simple brain of about 302 cells. Nonetheless it has a unique repertoire of behaviors and most importantly, it is amenable to genetic experimentation. The *aenorhabditis* genome consists of approximately 24,000 genes. (Blaxter 2003) The human body, comprised of over fifty trillion cells, contains only 1,500 more genes than the lowly, spineless, thousand-celled microscopic worm.

The fruit fly, another favored research subject, has 15,000 genes. (Blaxter 2003; Celniker, et al, 2002) So the profoundly more complicated fruit fly has 9,000 fewer genes than the more primitive *Caenorhabditis* worm. And when it comes to the question of mice and men, we might have to think more highly of them or less of ourselves; the results of parallel genome projects reveal that humans and rodents have roughly the same number of genes! (pp. 32–34)

◦

[The following] puts forth my nominee for the true brain that controls cellular life—the membrane. I believe that when you understand how the chemical and physical structure of the cell's membrane works, you'll start calling it, as I do, the magical membrane. Or alternatively, capitalizing on the fact that part of the word is a homophone for brain, I refer to it in my lectures as the magical mem-Brain. And when you couple your understanding of the magical membrane with an understanding of the exciting world of quantum physics that I'll present in the next chapter, you will also understand how wrong the tabloids were in 1953. The true secret of life does not lie in the famed double helix. The true secret of life lies in understanding the elegantly simple biological mechanisms of the magical membrane—the mechanisms by which your body translates environmental signals into behavior. (p. 45)

◦

So what structure in the prokaryotic cell provides its "intelligence"? The prokaryotes' cytoplasm has no evident organelles, such as the nucleus and mitochondria, that are found in more advanced, eukaryotic cells. The only organized cellular structure that can be considered a candidate for the prokaryote's brain is its cell membrane. (p. 46)

In 1985, I was living in a rented house on the spice-drenched Caribbean island of Grenada teaching at yet another "off-shore" medical school. It was 2 A.M., and I was up revisiting years of notes on the biology, chemistry, and physics of the cell membrane. At the time I was reviewing the mechanics of the membrane, trying to get a grasp of how it worked as an information processing system. That is when I experienced a moment of insight that transformed me, not into a crystal, but into a membrane-centered biologist who no longer had any excuses for messing up his life.

At that early morning hour, I was redefining my understanding of the structural organization of the membrane. Staring first with the lollipop-like phospholipid molecules and noting that they arranged in the membrane like regimented soldiers on parade in perfect alignment. By definition, a structure whose molecules are arranged in regular, repeated pattern is defined as a crystal. There are two fundamental types of crystals. The crystals that most people are familiar with are hard and resilient minerals like diamonds, rubies, and even salt. The second kind of crystal has a more fluid structure even though its molecules maintain an organized pattern. Familiar examples of *liquid crystals* include digital watch faces and laptop computer screens.

To better understand the nature of a liquid crystal, let's go back to those soldiers on parade. When the marching soldiers turn a corner, they maintain their regimented structure, even though they're moving individually. They're behaving like a flowing liquid, yet they do not lose their crystalline organization. The phospholipid molecules of the membrane behave in a similar fashion. Their fluid crystalline organization allows the membrane to dynamically alter its shape while maintaining its integrity, a necessary property for a supple membrane barrier. So in defining this character of the membrane I wrote: "The membrane is a liquid crystal." (pp. 59–60)

I had been trained as a nucleus-centered biologist as surely as Copernicus had been trained as an Earth-centered astronomer, so it was with a jolt that I realized that the gene-containing nucleus does not program the cell. Data is entered into the cell/computer via the membrane's receptors, which represent the cell's "keyboard." Receptors trigger the membrane's effector proteins, which act as the cell/computer's "Central Processing Unit" (CPU). The CPU effector proteins convert environmental information into the behavioral language of biology.

I realized in those early morning hours that even though biological thought is still preoccupied with genetic determinism, leading-edge cell research, which continues to unfold the mystery of the Magical Membrane in ever more complex detail, tells a far different story.

Over the years I gradually honed my presentation about the Magical Membrane and continued to refine it so that first-year medical students and lay people can understand it. I've also continued to update it with the latest research. In so doing, I've found much more receptive audiences among a wider range of people. (pp. 62–63)

From Sphere to Torus: A Topological View of the Metazoan Body Plan

Harald Jockusch and Andreas Dress

From the viewpoint of mathematical topology, membrane systems in intact living cells can be described as closed and orientable surfaces, i.e., as surfaces with two sides and no boundary lines so that an 'inside' and an 'outside' can be distinguished. Usually, biomembranes represent topological spheres, often one embedded in another one. Toroidal membranes are occasionally observed, e.g., in specialized structures of plant cells like the prolamellar body. Here, we propose that rules analogous to those that govern the topology of biomembranes hold for the epithelial cell sheets that cover anatomically external as well as internal surfaces of multicellular animals. We suggest to study the emergence of morphological complexity during metazoan development using concepts from mathematical topology, and propose experimental analyses of those topological transitions that appear to be relevant in development and evolution.

1. INTRODUCTION

While the morphological diversity of multicellular animals, the metazoa, seems overwhelming as exemplified by larval and adult forms of the marine fauna, the underlying body plans are nevertheless governed by rather few general principles. Boundaries between the outside world and the body as well as within the body are formed by cell layers, the epithelia, that are typically separated from the space filling tissue underneath by a layer of extracellular macromolecules, the basal lamina. It is the epithelia that determine the compartmentalization and thus the structural complexity of a metazoan. We show here that a general description of metazoan body plans can be given when

---

Based on a lecture held by H. J. at the Max Planck-Institute for Physics of Complex Systems, Dresden, 26 April 2001.
Reprinted from *Bulletin of Mathematical Biology* (2003) 65: 57–65; DOI: 10.1006/bulm.2002.0319.

one views the epithelia as surfaces,[1] and that its underlying principles are similar to those valid at the level of cellular membrane systems.

From the viewpoint of mathematical topology, membrane systems in intact living cells, from bacteria to eukaryotes including their cell organelles, can be described as *closed* and *orientable* surfaces (Schnepf, 1984; Lipowsky, 1991), i.e., as surfaces without boundary lines and with two distinguishable 'sides' that could be painted in a consistent fashion with two distinct colours. For example, a piece of paper is orientable, yet it is not closed having a boundary with the topology of a circle line; the *sphere*, i.e., the surface of a ball [round or deformed, Fig. 1(b)], and the *torus* i.e., of a tire or bagel [Fig. 1(c)], are both, closed and orientable, that is, they have two distinguishable sides and no boundary lines. In contrast, the *Möbius band* [Fig. 1(a)] is neither closed (having a boundary line with the topology of the circle line) nor orientable, and the *Klein bottle*, while being closed, is not orientable and cannot even be embedded in 3-D space without self-intersection which implies, of course, that biomembranes with the topology of a Klein bottle or a Möbius band cannot exist.

Usually, biomembranes form surfaces of roughly spherical shape, one often embedded in another one. Yet, surfaces with other topologies like that of a simple [Fig. 1(c)] or even a multiple torus (i.e., a 'pretzel' with two or even more 'holes') are formed by biomembranes in specialized structures of plant cells, e.g., the prolamellar body (Gunning and Steer, 1996). We propose that rules analogous to those that govern the topology of biomembranes hold for epithelia that separate the inner space of metazoan bodies from the outer world, provided one views these epithelia as surfaces.

The epithelia that cover the body surface, be it skin or intestine, consist of polarized cells that are all oriented in the same direction, thus defining at any point an outside facing, external space, and an inside facing the topological interior of the organism. Consequently, and in analogy to biomembranes, they form closed orientable surfaces of spherical or toroidal shape. This argument excludes cylindrical epithelia as long-lived structures because of their boundary lines though they can and, actually, must appear as short-lived intermediate stages—while even the transient formation of epithelial systems with the topology of the Möbius band or the Klein bottle is principally impossible.

During metazoan development, epithelia determine the increasing complexity of the body by actively changing their local curvature, e.g., during the formation of the primitive gut by invagination, leading to the gastrula stage (Odell *et al.*, 1981). Yet, according to the Gauss–Bonnet theorem [cf. Coxeter (1989)], an increase of curvature at some location has to be counterbalanced—as long as the overall topology of the epithelial surface

[1] We use the term 'surface' synonymous with the mathematically more stringent term '2-D manifold' (of points), NOT to designate the outer boundary of a 3-D body.

is not changed—by a corresponding decrease of curvature at another location. Such changes however—probably most of the time predated by drastic changes of local curvature—happen at certain stages of development. Thus, a formal analysis of the emergence of complexity during the process of development using the tools of topology and differential geometry should advance our understanding of organismic morphogenesis. Remarkably, even though Edelman (1988) coined the term 'topobiology' as a metaphor for the study of spatial aspects of development, such an analysis has, to our knowledge, not yet been proposed in the literature [see Chaplain et al. (1999) for a recent review].

2. TOPOLOGICAL TRANSITIONS IN METAZOAN DEVELOPMENT

The true multicellular animals, the eumetazoa, include the *diploblastic Ctenophora* and *Cnidaria* (collectively called *Coelenterata*) to which polyps and jellyfish (medusae) belong, as well as the *triploblastic Bilateria* which comprise invertebrate and vertebrate animals, from 'worm' to man [cf. Westheide and Rieger (1996)]. The body plan of an individual polyp or medusa [Fig. 1(d), upper right] is topologically a sphere, derived from the gastrula-like larva [Fig. l(d), left]. However, during budding, secondary body openings are formed independently of the initial mouth. The lumen of the gut, representing the exterior in relation to the animal, may extend and branch in complex ways without affecting the spherical topology of the body plan, but secondary mouth openings connected by a common gut transform it into a (multiple) toroid [Fig. l (d), lower right]. The formation of a torus is also a regular acquisition of the bilaterian metazoa [Fig. 1(e)] except for flatworms [*Plathelminthes*, Fig. 1(g)].

Whereas gastrulation typically involves drastic changes of epithelial curvature [Fig. 1(e)] and creates new cell-to-cell neighbourhoods [and therefore has been called 'truly the most important time in your life' by L. Wolpert (1983); cited in Gilbert (1991)], the resulting gastrula retains the spherical topology of the blastula. In contrast, the formation of a second body opening is a topological 'breakthrough' (often also physically, in the literal sense of the word) because it transforms the spherical surface of the gastrula into one of toroidal shape [Fig. l(d) and l(e), right].

In several groups of animals, including the cnidarian polyp *Hydra*, the mechanism of gastrulation is different from the invagination described above and may involve separation of cells from the outer layer of the spherical blastula into the inner space where a new cell layer is formed, resulting in two spheres, one embedded within the other [cf. Fiorini (1992)]. In order to form a primitive mouth, a fusion of the two spheres and a subsequent 'breakthrough' is required to generate a spherical gastrula. However, gastrulation by invagination is considered the original, evolutionarily more primitive mechanism.

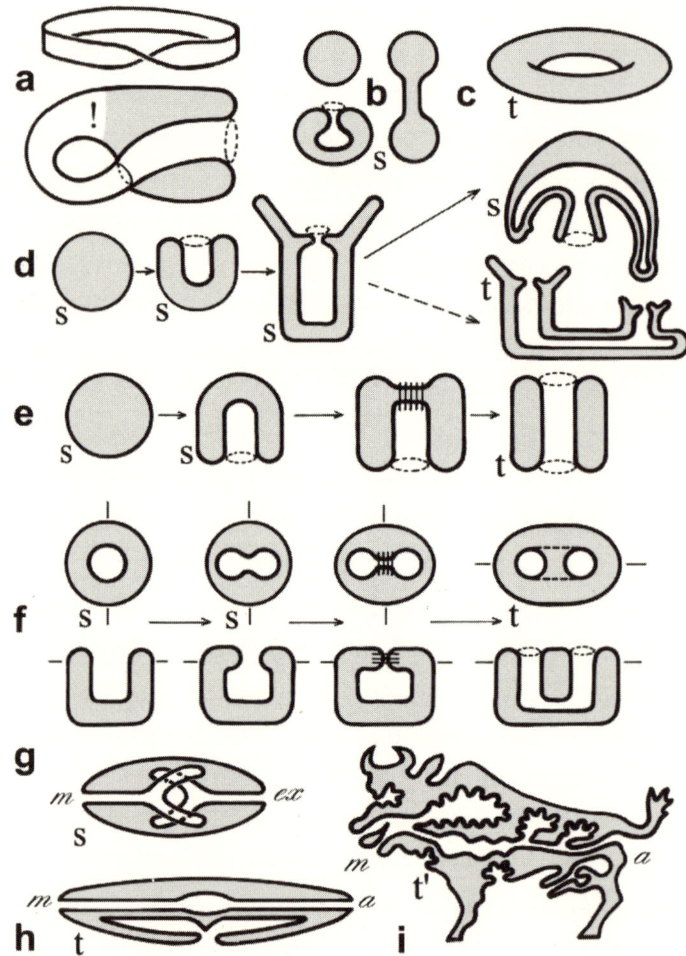

Figure 1. The Möbius band, closed two-dimensional surfaces, and metazoan body plans. *s*, sphere; *t*, torus; *t0*, toroid; shaded areas, 'interior.' (a) Nonorientable surfaces, the nonclosed Möbius band (up) and the closed 'Klein bottle' (section, down). (b), (c) Two-dimensional orientable surfaces; (b), spheres of different shapes (sphere proper, dumb-bell, 'inverted dumb-bell'), sections; (c) torus ('tire'), as seen from 'outside.' (d)–(h) Metazoan development and body plans (sections). Bold lines represent epithelia, 'stitches' indicate fusion of epithelia. *m*, mouth; *a*, anus. (d) Single coelenterates, polyp or medusa, are topological spheres; colonial and budding stages are (multiple) toroids, due to secondary mouth openings. (e)-(i) Bilaterian development and body plans: except for flatworms [(g); *ex*, excretion organ], there is a transition from sphere to torus during the formation of the second opening of the primitive gut (anus in protostomia, definitive mouth in deuterostomia). (f) Alternative formation of second gut opening in certain protostomians (see text for explanation). Torus-like body plans of (h) a round worm (nematode) and (i) a vertebrate (bull, ital., span. *toro*). In (h), epidermal invaginations schematically indicate lacrimal (head), sweat and sebaceous glands (in a female, mammary glands would have to be added). Evaginations of the gut represent epithelial linings of lungs, pancreas and liver.

In the deuterostomes (e.g., sea urchins and vertebrates) where the blastopore becomes the anus, the definitive mouth is generated by a coordinated formation of a bounded hole involving the endodermal and the overlying ectodermal epithelial cell layers in a process that could be described as 'self-wounding and healing' [Fig. 1(e), right]. In many protostomes, it is the anus that is formed by a similar mechanism. An alternative mechanism is observed in certain groups of protostomia [like the *Annelida*; cf. Westheide and Rieger (1996), Arendt et al. (2401)]: the blastopore elongates to form a double-walled slit and fuses in the middle, thus producing two gut openings [Fig. 1(f)]. Yet, in all of these cases, there is an increase of *topological* complexity to be discussed below.

3. SURFACES, THEIR GENUS, AND THEIR EULER CHARACTERISTIC

The topology of a closed connected and oriented surface *M* (where 'connected' means 'in one piece,' and not in two separate pieces like e.g., two distinct spheres—whether one is embedded into the other or not) is completely determined by a single integer, its Euler characteristic _(*M*). This number can be read off easily from any *triangulation* of *M*. It is the alternating sum of the number of vertices, edges, and faces appearing in that triangulation, i.e., we have _(*M*) = *V*—*E* + *F*, where *V*, *E*, and *F* are the numbers of vertices, edges and faces, respectively. It is closely related to the genus ('Geschlecht') g(*M*) of *M*, the number of 'handles' (or 'holes' created by handles), i.e., the identity _(*M*) = 2—2g(*M*) holds for every closed, connected and oriented surface M—in particular, _(*M*) is always an even number and never larger than 2 for such surfaces. Thus, spheres have genus 0 and Euler characteristic 2 while a (simple) torus has genus 1 and Euler characteristic 0, and the genus of all higher-order toroids is _2 and their Euler characteristic is negative. Furthermore, the above definition _(*M*) = *V*—*E* + *F* also works for nonconnected surfaces, and also for surfaces with a boundary, i.e., it is a *topological* invariant for all of these surfaces and never depends, e.g., on the actual triangulation used for computing it (while it is, of course, not sufficient to characterize the topology of these more general surfaces completely). Note that the Euler characteristic of the union of two separate, disconnected surfaces is the sum of _'s of the two individual surfaces.

4. SURGERY THEORY OF SURFACES

In *surgery theory* (Stewart, 1995; Devlin, 1999), a branch of topology, one studies simple geometric operations that can be used to transform one (not necessarily connected, closed, or oriented) surface into another. In two dimensions, the standard surgery operation consists of 'cutting out' two separate 'lids' from a surface *M*, thus producing a new 'wounded' surface *N* with two circular holes surrounded by two circular boundary lines, and then 'gluing' the two circular 'rims' of a hollow cylinder to those two circular boundary lines of *N* to 'heal its wounds.'

If we imagine *M* as being the surface of a hollow 3-D body, this can be done 'from its inside,' so that the 'lumen' of the cylinder becomes a tunnel-like part of outside space, or 'from its outside,' so that the cylinder forms a 'handle' and its 'lumen' becomes part of the 'inside' of the resulting object. In both cases, the Euler characteristic is reduced by 2.

One can also reverse this process provided a 'handle' or a 'tunnel' has been specified: one can cut out the handle and glue two circular 'discs' into the resulting two holes, or cut out the tunnel's surface and fill the resulting two holes with two circular discs. In both cases, the Euler characteristic is increased by 2.

Thus, starting for instance with a (deformed) sphere *S* [as in Fig. 2(b)] and replacing a cylindrical part of this surface by two lids, we increase the Euler characteristic by 2. So, this will necessarily result in two disconnected spheres because there is no surface with Euler characteristic 4 = 2_(*S*). The two resulting spheres will be separated in space when the cylinder is the central part of a dumbbell [Fig. 1(b), right]. However, when the cylinder points inwards [forming an 'inverted dumb-bell,' Fig. 1(b), lower left], one of the spheres will be engulfed by the other.

5. "SURGERY OPERATIONS" IN METAZOAN DEVELOPMENT

Although cylinders and discs and 'cutting' and 'gluing' of 'handles' do not actually occur in living organisms, these mathematical metaphors are helpful concepts to clarify topological transitions of membrane and epithelial systems. The process of disconnecting a sphere into two spheres by cutting out a cylinder that is the central part of a dumb-bell is observed—on the level of biomembranes—in cell division and budding. The same process, yet with the cylinder pointing inwards, is observed in endocytosis at the cellular level and in neural-tube formation in vertebrate development.

The reversion of such a mechanism, i.e., the fusion of an inner sphere with an outer sphere is observed in cellular membranes and called exocytosis. As mentioned above, this transition is observed for epithelia during gastrulation in some diploblastic animals. Topologically, this process reduces the Euler characteristic by 2, from the sum 4 of the two disconnected spheres to the number 2 for a single sphere.

Mechanisms for torus formation have probably evolved several times independently, in both coelenterata and bilateria. Repeated addition of 'handles' or creation of 'holes' has led to double, triple, and even higher-order toroids (g = 2, 3, …; Euler characteristic

_ = –2, –4, …). Vertebrates are at least triple tori because their nostrils connect to the gastrointestinal tract, whereas ears, the urogenital system, and the sweat, sebaceous, and mammary glands in our skin and the epithelial linings of organs that are appendices to the intestine, like liver and pancreas, as well as the bronchia and the lung are surface invaginations and hence topologically irrelevant [Fig. l(g)–1(1)]. In triploblastic animals, the internal mesenchyme may form epithelia *de novo*. This is how the circulatory system of vertebrates arises. The surface of the cardiovascular system constitutes, due to extensive anastomosis of blood capillaries, a high order toroid.

Our considerations regarding epithelia as closed orientable surfaces are valid independently of any controversies regarding the polarity of body axes, i.e., whether the mouth or the anus of higher metazoa is homologous to the single body opening of *Cnidaria* (Arendt *et al.*, 2001), or whether the 'head' or the 'foot' is homologous to the 'anterior' pole of the body axis of *Bilateria* (Meinhardt, 2002).

Surprisingly, nearly nothing is known about the mechanism of the formation of the secondary body openings in either *Coelenterata* or *Bilateria.* Since developmental processes, in particular species differences in development, are to a large extent genetically controlled, it should be possible to identify mutations affecting the sphere ! torus (g = 0 ! g = 1) transition. As such mutations are expected to be lethal, one would need an early-embryo screen to detect them. The problem will be to find an organism in which the 'breakthrough' starts (ideally) from a monolayered gastrula (like the chordate *Branchiostoma*, the lancet) and which, at the same time, is tractable by developmental genetics (like the zebrafish). In addition to defects of the topological transition process, one might find 'craftsman genes' involved in the physical realization of the topological program, and these may be similar to those active during wound healing (coding e.g., for extracellular matrix-degrading enzymes). In contrast, the hypothetical gene(s) searched for would provide the initiation and program of torus formation, i.e., the gene functions acquired during the evolutionary transition from simple 'spherical' coelenterata to colonial forms, and from the primitive flatworms to the torus-shaped 'higher' *Bilateria.*

6. CONCLUSIONS AND OUTLOOK

What would be the expected characteristic of a 'breakthrough gene'? Given the complex orchestration required for epithelial fusion and breakthrough, a 'master' regulatory gene might be a likely candidate. Furthermore, most researchers would expect a site-specifically acting gene to be site-specifically expressed (although this need not necessarily be so), i.e., the gene might be expressed at, or close to, the site where the breakthrough is expected to take place. Alternatively, in the case of a negative regulator, this site might be specifically spared of the expression of the breakthrough regulator gene.

In both cases, an 'upstream' mechanism would be required to regulate the local expression of the regulator. In *Cnidaria* which are assumed to have conserved evolutionary old mechanisms for morphogenesis, specific expression of the genes *brachyury* (Technau and Bode, 1999), *Wnt* (Hobmayer *et al.*, 2000) and *goosecoid* (Arendt *et al.*, 2001) has been observed near the ends of the gut. Of these, *Wnt* might be a particularly interesting candidate for a 'breakthrough gene' [cf. Meinhardt (2002)]. However, a direct involvement in the breakthrough mechanism has in no case been experimentally demonstrated or even inferred from indirect evidence. Surprisingly, even an accurate description of the topologically fundamental process of 'breakthrough' is, to our knowledge, not available in the literature. Thus, the search for the topologically relevant 'breakthrough gene(s)' might disclose fundamental transitions and novel mechanisms.

Acknowledgments

We are particularly grateful to Drs Gary Freeman (University of Texas, Houston), Eberhard Schnepf, Jochen Wittbrodt (both Heidelberg), Hans Meinhardt (Tübingen), David Epstein (Warwick, U.K.) and Peter Heimann (Bielefeld University) for discussions and help.

In a Letter to the Editor (*Bulletin of Mathematical Biology* (2004) 66: 1455; DOI: 10.1016/j.bulm.2004.06.002), authors Harald Jockusch and Andreas Dress acknowledge the work of Presnov and colleagues (Maresin and Presnov, 1985; Presnov et al., 1988) in which the basic ideas of the topology of body plans during development and evolution have already been proposed.

References

Arendt, D., U. Technau and J. Wittbrodt (2001). Evolution of the bilaterian larval foregut. *Nature* 409, 81–85.

Chaplain, M., G. Singh and J. McLachlan (1999). *On Growth and Form: Spatio-temporal Pattern Formation in Biology*, J. Wiley & Son.

Coxeter, H. S. M. (1989). *Introduction to Geometry*, New York: Wiley.

Devlin, K. (1999). *Mathematics: The New Golden Age*, New York: Columbia University Press.

Edelmann, G. M. (1988). *Topobiology. An Introduction to Molecular Embryology*, New York: Basic Books.

Fiorini, P. (1992). *Allgemeine und Vergleichende Embryologie der Tiere*, Berlin: Springer.

Gilbert, S. F. (1991). *Developmental Biology*, Sinauer Associates.

Gunning, B. and M. Steer (1996). *Plant Cell Biology*, Sudbury, MA: Jones & Bartlett.

Hobmayer, B., F. Rentzsch, K. Kuhn, C. M. Happel, C. C. von Laue, P. Snyder, U. Rothbacher and T. W. Holstein (2000). WNT signalling molecules act in axis formation in the diploblastic metazoan hydra. *Nature* 407, 186–189.

Lipowsky, R. (1991). The conformation of membranes. *Nature* 349, 475–481.

Meinhardt, H. (2002). The radial-symmetric hydra and the evolution of the bilateral body plan: an old body became a young brain. *Bioessays* 24, 185–19 1.

Odell, G., G. Oster, P. Alberch and B. Burnside (1981). The mechanical basis of morphogenesis. I. Epithelial folding and invagination. *Dev. Biol.* 84, 446–462.

Schnepf, E. (1984). The cytological viewpoint of functional compartmentation, in *Compartments in Algal Cells and their Interaction*, W. Wiessner, D. Robinson and R. Starr (Eds), Berlin: Springer.

Stewart, I. (1995). *Concepts of Modern Mathematics*, New York: Dover Publications.

Technau, U. and H. R. Bode (1999). HyBra1, a brachyury homologue, acts during head formation in hydra. *Development* 126, 999–1010.

Westheide, W. and R. Rieger (1996). *Spezielle Zoologie: Einzeller und Wirbellose Tiere*, Stuttgart: G. Fischer.

A Gene Regulatory Network Subcircuit Drives a Dynamic Pattern of Gene Expression

Joel Smith, Christina Theodoris, and Eric H. Davidson

ABSTRACT

Early specification of endomesodermal territories in the sea urchin embryo depends on a moving torus of regulatory gene expression. We show how this dynamic patterning function is encoded in a gene regulatory network (GRN) subcircuit that includes the *otx, wnt8,* and *blimp1* genes, the cis-regulatory control systems of which have all been experimentally defined. A cis-regulatory reconstruction experiment revealed that *blimp1* autorepression accounts for progressive extinction of expression in the center of the torus, whereas its outward expansion follows reception of the Wnt8 ligand by adjacent cells. GRN circuitry thus controls not only static spatial assignment in development but also dynamic regulatory patterning.

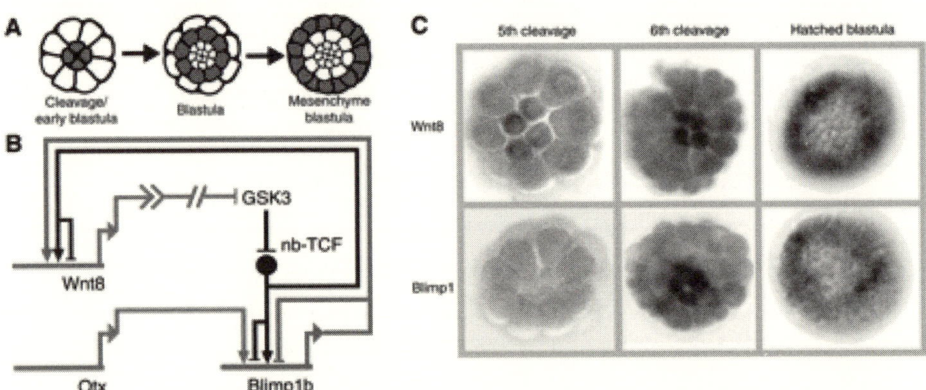

Fig. 1. Moving-torus gene expression pattern. (**A**) Representation of expression pattern of *blimp1* or *wnt8* genes (red). The innermost cells are skeletogenic micromeres; the red ring in the second drawing shows mesoderm cells (prospective secondary mesenchyme); the outer ring is definitive endoderm. Expression of the *blimp1* gene begins in the micromeres around 6 hours after fertilization and appears in the adjacent tier of mesodermal cells by 12 hours. Soon after, expression disappears from the micromeres. By 18 hours, expression of *blimp1* begins in the adjacent presumptive endoderm lineage and disappears from the mesodermal cells.

Excerpted from *Science* (2007) 318: 794–797; DOI: 10.1126/science.1146524. Reprinted with permission from AAAS.

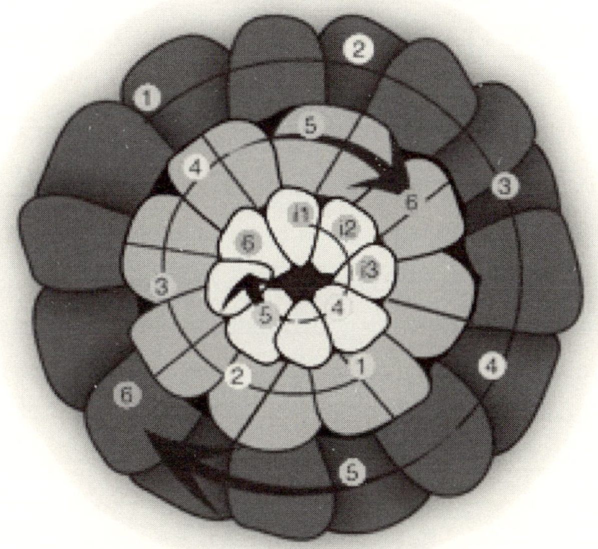

Initial steps (inner tier only)

1. Initial State: Groucho-Tcl represses zygotic *blimp1* and *wnt8* expression
2. Initial Input: Maternal anisotropy stabilizes nß-catenin at vegetal pole
3. Nuclear ß-catenin + maternal Blimp initiate *wnt8* expression; nß-catenin + Otx drive zygotic *blimp1* (to step 4)

Initial steps (middle and outer tiers)

1. Wnt8 ligand diffuses from inner tier to middle tier or middle tier to outer tier
2. nß-catenin + Otx drive *blimp1* expression
3. βeta-catenin + Blimp1 drive *wnt8* expression (to step 4)

Common to all tiers

4. Wnt8 signaling: among single-generating cells increases nß-catenin: ligand diffuses to neighboring tier (to step 1 in next tier)
5. Blimp1 represses self
6. In absence of Blimp, *wnt8* turns off

Fig. 3. Summary of mechanism by which the dynamic, concentrically expanding torus of *wnt8* and *blimp1* expression is generated. The drawing shows a seventh-cleavage embryo viewed from the vegetal pole to illustrate the radial concentric organization. However, the events indicated in the numbered key begin at fourth cleavage and extend out to mesenchyme blastula stage (18 to 24 hours).

Topological Patterns in Metazoan Evolution and Development

Valeria Isaeva, Eugene Presnov, and Alexey Chernyshev

ABSTRACT

Topological patterns in the development and evolution of metazoa, from sponges to chordates, are considered by means of previously elaborated methodology, with the genus of the surface used as a topological invariant. By this means metazoan morphogenesis may be represented as topological modification(s) of the epithelial surfaces of an animal body. The animal body surface is an interface between an organism and its environment, and topological transformations of the body surface during metazoan development and evolution results in better distribution of flows to and from the external medium, regarded as the source of nutrients and oxygen and the sink of excreta, so ensuring greater metabolic intensity. In sponges and some Cnidaria the increase of this genus up to high values and the shaping of topologically complicated fractal-like systems are evident. In most Bilateria a stable topological pattern with a through digestive tube is formed, and the subsequent topological complications of other systems can also appear. The present paper provides a topological interpretation of some developmental events through the use of well-known mathematical concepts and theorems; the relationship between local and global orders in metazoan development, i.e., between local morphogenetic processes and integral developmental patterns, is established. Thus, this methodology reveals a "topological imperative": A certain set of topological rules that constrains and directs biological morphogenesis.

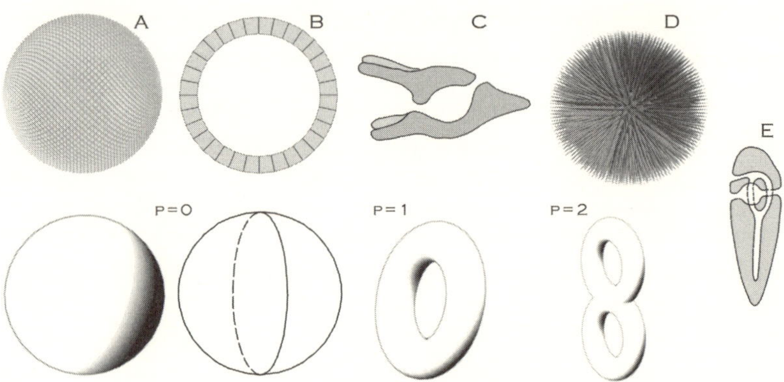

Fig. 6. Vector fields on surfaces: field singularities on a sphere, smooth field on a torus.

Excerpted from *Bulletin of Mathematical Biology* (2006) 68: 2053–2067; DOI: 10.1007/S1158-006-9063-2.

Broken Symmetries and Biological Patterns

Ian Stewart

10.1 INTRODUCTION

The search for pattern in the natural world is one of the key motivating forces of science. Powerful techniques including deep mathematical theories, computer simulations and sophisticated experiments using complex apparatus have been developed in order to analyse patterns and how they are formed. Mostly, these methods see pattern as an accidental side effect, emerging from reductionist rules. It would represent a major step forwards if we could somehow develop a theory of pattern formation that made the role of the patterns themselves central. An example would be a theory of hurricanes in which the key conceptual element is a huge rotating spiral of gas and vapour—rather than one in which the key conceptual element is a system of equations representing the interactions of tiny packets of gas or vapour and from which the awesome reality of hurricanes emerges almost by accident.

Biology has a particular need for such a synthesis. Mainstream biology has largely concerned itself with *things*—first organisms, then cells, now the molecular structure of hormones, proteins, RNA and DNA. Because of this historical development, most biologists prefer reductionist explanations of nature, in which the forms and patterns of particular things are consequences of other things that can be found inside them. An example of such thinking is the human genome project, which was largely promoted on the assumption that DNA contains all of the 'information' needed to determine an organism, so that sequencing human DNA will tell us everything we wish to know about human biology. This view is both profound and naive: profound in that the molecular nature of genetics represents the biggest breakthrough in biological understanding since the invention of the microscope and naive in that it assumes that it is the ultimate breakthrough, and that all else will be a kind of gloss upon DNA. Recent puzzlement that the human genome contains only 35000 or so genes, in contrast to the more than 100000 proteins in human cells, points to a conceptual gap. This is reinforced by the recent proliferation of 'omes'—proteome, legome, transcriptome—whose avowed intention is to bridge the gaps between DNA, cells and organisms.

Excerpted from *On Growth, Form and Computers,* Edited by Sanjeev Kumar and Peter J. Bentley, Elsevier Academic Press, 2003, pp.181–183.

The idea that the explanation of biological patterns should have a mathematical element is not new. Wilhelm His (see Gould, 1987) attempted to counter Ernst Haeckel's view that 'ontogeny recapitulates phylogeny'—that embryos 'climb their own family tree' by passing through stages of development that represent evolutionary ancestors. His argued that a lot of embryological development is the result of purely mechanical laws and he demonstrated it by cutting rubber tubes in various ways and folding them up to get forms very similar to those arising in embryos. His was ridiculed: Haeckel talked of a 'sartorial theory of embryology' and of the 'ragbag' or 'rubber-tube' theory. Haeckel's own theory has turned out to be totally misconceived. His's theory may yet be rehabilitated, though in a radically different form, because many features of development include mechanical effects (Murray, 1989; Goodwin and Brière 1992; Goodwin 1994).

One of the best known early advocates of biomathematics was D'Arcy Thompson (1942), who emphasized common mathematical patterns in the inorganic and organic worlds. Conrad Waddington was less overtly mathematical, but invented concepts such as 'canalization'—a kind of inherent stability of wild-type genetics, which he explained using quasi-mathematical metaphors (Waddington, 1975). Alan Turing (1952) attempted to provide mathematical foundations for the development of form and pattern in living creatures, by specifying mathematical equations obeyed by substances called 'morphogens' which, he believed, laid down a pre-pattern for subsequent development to build on. In the 1960s the French mathematician René Thom (1975) invented 'catastrophe theory,' which attempted to classify sudden changes in geometric terms—inspired to a great extent by Waddington's attempts to find a dynamic formalism for biological development. Many specific models based on his ideas were proposed by Zeeman (1977), but on the whole these did not take root among biologists. The latest episode in this long-running saga is Stuart Kauffman's work on complex adaptive systems and autonomous agents, expanding into the 'space of the adjacent possible' as rapidly as they can manage (Kauffman, 1993, 2000). There is a deep common theme: in place of molecular information and more or less arbitrary 'instructions,' all of these visionaries have emphasized geometry and dynamics.

The enormous success of the approach unleashed by Crick and Watson's discovery of the DNA double-helix is undeniable and its importance should not be underestimated. However, there is a serious danger that future generations of biologists, brought up within that molecular paradigm, will ignore the associated dynamical processes of development and behaviour, believing them all to be reducible to DNA. However, DNA 'instructions' cannot be the whole developmental story, for a number of reasons, among them the following (Lewontin, 1992; Cohen and Stewart, 1994; Stewart, 1998):

- Evolution is opportunist and will exploit any structures and processes that come 'for free' as a consequence of the laws of physics; therefore these structures do not *need* complete specification within DNA and therefore they are unlikely to be so specified.
- Organisms transmit far more than just genetic material to their developing offspring. They provide a context and an environment for development—a fully functioning support system within which the genes can begin to function. Parent organisms provide materials, spatial geometry and temporal organization.
- The amount of information in DNA (i.e. the number of bases, 300 billion in humans) is insufficient to define the structure of an adult organism (e.g. the 10^{12} connections to be made in the human brain). The best it can do is define *processes* that generate the required structures. But then those processes must, to a great extent, be 'free-running.'

If these structures and processes operated transparently, then it would be reasonable to consign them to an unspecified physical 'default' background, as many biologists do. However, many crucial aspects of the biology are concealed within the subtleties of physicochemical processes and it is essential to understand how these processes work. For example, DNA specifies the amino acid sequence of a protein, but physicochemical processes prescribe what shape it folds into. Protein function is determined almost entirely by the shape, so the effect of the DNA sequence is cryptic. Most amino acid chains fold into a bewildering variety of shapes and it is virtually impossible (NP-hard, in the computing jargon) to compute the lowest-energy shape from the sequence. It appears that evolution may have favoured sequences that naturally fold in a robust manner. So here we see a deep complicity (Cohen and Stewart, 1994; Stewart and Cohen, 1997) between DNA, physics and evolution, which cannot usefully be referred back just to DNA.

Ironically, while many biologists have been rejecting the model of development as a dynamic process, seeing it as something more akin to a knitting pattern written in DNA and woven in proteins, the mathematical mainstream has been moving in precisely the opposite direction. Modern mathematical theories of dynamics and pattern-formation emphasize qualitative features and processes, rather than reductionist fine details (Langton, 1986; Stewart, 1989; Stewart and Golubitsky, 1992). Such theories may constitute a first step towards the programme of understanding pattern formation in terms of the patterns themselves—what I have elsewhere (Stewart, 1995) called 'morphomatics.'

There is a growing mismatch between biologists' perceptions of mathematics (classical, rigid, deterministic, simple, quantitative, linear, rigorous) and the actuality (modern, flexible, adaptive, complex, qualitative, non-linear, speculative).

Self-Organization vs. Gene Regulation

Kathy Hall, Richard Milner, and Stuart Pivar

ABSTRACT

For the last twenty-five years, researchers have assumed that morphology is a direct expression of gene activity. Many scientists now think that form is not encoded in the genes but lies in innate patterns governed by the laws of physics and chemistry. Here we review the current literature that argues against genetic regulation of form and the concurrent view that natural selection does not dictate phenotypic form.

KEYWORDS

evolution; development; embryology; biomechanics; topobiology

Plant and animal life develop from a single fertilized egg, often yielding millions of embryonic cells in a dazzling array of organization, producing body forms from *Volvox* to flies, worms, insects and man. With the discovery of the structure of the DNA molecule in 1953, the vast majority of biological scientists accepted genetics as the key to understanding the fundamental phenomena of biology, including development and evolution. It became almost an article of faith that determining the exact structure of the genome of humans and other species would reveal how the body is assembled (Cowin 2004). The genome was called the "blueprint of life," the "holy grail" of life, with all details of form hidden within. The idea that genes would solve the problem of morphogenesis was accepted as dogma by biologists (Lewontin 2000; Zimmer 2001; Ridley 1999). It followed that natural selection of random changes in its structure was the phenomenon that shaped species during evolution (Ingber 1998).

Rather than launching biology into a new era of understanding, the complete decoding of the human genome announced in December 2000 augured a crisis in biology. Long thought to comprise over a hundred thousand genes, the human genome was found to contain merely thirty thousand. The decoding of the mouse genome in December 2002, also numbered thirty thousand, the same number as man, and of these, only three hundred differed from the human genes, a mere 1%. The 1,000 cell roundworm has 19,500 genes and the corn plant 40,000, 10,000 more than human (Ast 2005).

This small number of genes and lack of species diversity in the DNA suggested an insufficiency of information to distinguish the body form of man from mouse. Scientists are trying to account for this paucity of genetic information by cutting and pasting. How can 40,000 genes define the form and life of corn and 30,000 genes account for the complexity of humans? How can man be only 300 proteins different than a mouse? The only answer has come from gene splicing and perhaps one gene generating many forms of mRNA (Ast 2005). Surely after the millions spent in developmental biology grants, there is more to go on than this.

The question of genetics and form was not the only major question unanswered in biology. Gerald Edelman in *Topobiology* (1988) listed the related problem of the evolution of form. He stated, "As usual in fundamental matters, a mystery remains. This mystery is embedded in the problem of morphologic evolution, the process that gives animals and their organs their functional shapes ... what is the difference between a discrete genetic change that converts normal hemoglobin to sickle-cell hemoglobin with morbid consequences, and a genetic change that alters the shape and size of the whole animal or even of its nose?"

Even before the genome was fully decoded, cracks in the dogma began appearing. As Werner Müller put it in the 1996 edition of *Developmental Biology*, a standard college text: "It is commonly stated that the genome incorporates a Bauplan, an architectural plan, or blueprint of the body. Actually this is not the case; the genome is not a sketch or design of the finished body. The informational capacity of DNA is simply too low to store the blueprint of the final pattern of the organism. For example, the detailed design of the hundred trillion to one quadrillion synaptic contacts of the brain alone would greatly exceed the genomic memory ... We do not understand in detail how a developing organism is created on the basis of such minimal information, or how many organisms are able to regenerate lost structures."

Ultimately, it seems, knowledge of the genome shed little light on the study of morphogenesis. In the December 6, 2002 issue of *Science*, Ray Keller of the University of Virginia began an article on "Shaping the Vertebrate Body Plan by Polarized Embryonic Cell Movements" with the admission: "How the body plan is shaped from a cohesive aggregate of individual cells during embryogenesis is an enduring mystery." There is no confirmation that the genome is the code for phyletic form. So great has been the success of biology in medicine, especially genetic engineering, that it is amazing to realize biology is a branch of science which lacks a coherent fundamental theory of development.

Richard Lewontin, an evolutionary biologist, wrote in his preface to Susan Oyama's *The Ontogeny of Information* (2000), "Already in 1982, at the Cambridge University symposium commemorating the one hundredth anniversary of Darwin's death, Sydney

Brenner in his keynote address had declared that given the complete DNA sequence of an organism and a large enough computer it would soon be possible to compute the organism. By 1985, the genetic boa constrictor had totally enfolded embryology in its helical coils. Now, nearly fifteen years later, the helpless victim has disappeared down the molecular maw and is slowly being digested. In the end, of course, the result will be what is to be expected from such a process: a slight enlargement of the body of the consumer and the production of a large amount of fecal matter."

It is no wonder that biologists now question the role of genes in phyletic form. Although there is presently an actual catalogue of all the genes in organisms such as *Caenorhabditis elegans, Drosophila melanogaster* and *Arabidopsis thaliana*, the understanding of genetics has thus far not resolved the issues relating genotype to phenotype (Larsen 2003). We cannot relate the genome to the determination of morphology and form. There is no "Jurassic Park." The dinosaur cannot be re-created from a handful of DNA trapped in amber. "The form and behavior of a developing individual in any given species is not directly specified in the DNA saying 'make me an elephant'" (Dover 2000).

Man having the same amount of DNA as mouse is reminiscent of the work in the early nineteenth century, Étienne Geoffroy Saint-Hilaire declared that there is only one animal and Goethe concluded there was only one plant. Geoffroy was convinced that vertebrates and invertebrates with external exoskeletons were one and the same thing. "Their different appearance was due to specific distortions in the processes of growth and development that made one look like an inside-out version of the other ... This led Geoffroy to a belief in 'principle of connections' between basic body parts. These connections are capable of twisting and turning in many different ways, leading to bizarre configurations that look very unlike each other. Yet, unpick the developmental distortions of the connections in any given animal and the basic proto-animal is still there" (Dover 2000). Saint-Hilaire believed that insects are vertebrates. 'Look at the tortoise,' he said, 'and imagine that not only the ribs have fused into a half ball, but that the spine itself has joined the dome.' Thus to Saint-Hilaire an insect was a vertebrate animal which, having radically shifted the positions of all its organs and supporting systems, had ended up living inside its own spine" (Le Guyader 2004). An insect was nothing more than a vertebrate turned inside out.

Although Saint-Hilaire's ideas were shelved for a couple of hundred years, biologists are beginning once again to look at the possibility of one common animal and one plant. A walk through the nearest museum shows common structure. Nature has common assembly rules, which are implied by their reoccurrence, at scales from the molecular to the macroscopic of certain patterns, such as spirals, pentagons and triangulated forms. These patterns as shown by D'Arcy Thompson in his 1917 *On Growth and Form,* appear in structures as diverse as viruses, plankton and humans, made up of the same building blocks of carbon, hydrogen, oxygen, nitrogen and phosphorus, all arranged in three dimensional space (Ingber 1998; Thompson 1917; Gordon 1966). "The phenomenon, in which components join together to form larger, stable structures having new properties that could not have been predicted from the characteristics of their individual parts, is known as self-assembly and is observed at many scales in nature from the self-assembly of organelles into cells to the self-assembly of cells into tissues and tissues into organs" (Ingber, 1998).

SELF-ASSEMBLY

Stuart Kauffman's *The Origins of Order* (1993) says that self-organization plays an important role in the emergence of life and may play as fundamental a role in shaping life's subsequent evolution, as does the Darwinian process of natural selection. "No systematic effort has been made to incorporate the concept of self-organization into evolutionary theory. The construction requirements which permit complex systems to adapt remain poorly understood." So far no one has detailed just how nature's patterns and symmetries are directly coded for in the genome (Miodownik 2003). "The patterns of life must arise without detailed blueprints, just as an exquisitely symmetrical snowflake emerges from the random collisions of water molecules in moist air" (Cho 2004).

There are many examples of natural phenomena that are reproducible without being encoded by any genetic program. Transmission of structure and form to progeny does not have to be dependent on the transmission of genetic material (Minelli 2003). "There are default morphologies, much like an individual default fingerprint that is accurately reproduced over and over and dependent on only non-genetic cellular information" (Fusco 2000). Cells grow and divide using the template of a pre-existing cell. One centriole produces two using a pre-existing template. There is a general property of cells that determines its axis, independent of any gene's instructions. For instance yeast cells repeatedly polarize and will continually bud from a particular pole. The polarization is due to a membrane protein called Rax2 that takes a fixed memorized multigenerational position in the cell's cortex (Müller 1996; Chen 2000).

Cytoplasmic memory is well known in cells, in particular in that of ciliates which repeat, generation after generation, perfectly patterned rows of cilia without any direct intervention from the genome (Tartar 1961). In the 1960's, Tracy Sonnebom discovered that he could transplant bands of cilia on a Paramecium by slicing off chunks of membrane with attached cilia and rotating the direction of the beating cilia. Paramecium divides by splitting in two and each of the daughter cells of Sonnebom's surgically altered Paramecium had cilia that beat in the opposite direction. Furthermore this surgically altered graft was carried into succeeding generations (Sonnebom 1965). No change in the DNA occurred, but each daughter cell now carried the pattern of the misplaced cilia. Another example cited by Minelli (2003) is the self-assembly of the multicellular alga *Hydrodyction*. This alga has a three-dimensional lattice structure made of many cells, which also serve as a non-genetic template for the three-dimensional pattern of the next generation (Bonner 1993). Minelli said there are many examples of organisms transmitting major morphological traits or developmental patterns by non-genetic means, but he questioned whether this was possible in metazoan animals due to the size limits of the egg and the amount of information necessary for the patterning of the adult body plan.

Self-assembly or self-organization is thought to occur by the laws of physics, chemistry and mathematics. "A … mechanism for the morphogenetic event of pattern formation in tissue development is mechanical instability, or buckling. The first example of mechanical instability encountered in mechanics is the column in which only small deflections are considered. If the column is simply supported at both ends, it will deflect from its straight shape into the shape of a half-sine wave when the compressive axial load acting on the column reaches a critical value called the critical load. There are other shapes that it will take as the compressive load is further increased" (Cowin 2000).

Mathematician and biologist Ian Stewart writes that circumferential bands forming patterns in the egg could result from buckling under compression failure (Stewart 1998)). According to Stewart, exactly the same buckling happens if uniform forces directed toward its center compress any kind of spherical shell such as a ping-pong ball.

"Induce instability in a spherical shell and allow it to buckle spontaneously into a circular dent, collapsing in on itself. This scenario comes free of charge to frog biology. Frog genetics may trigger the instability and if necessary, it can evolve to modify the precise buckled form, but it doesn't need to tell the blastula how to buckle. Similarly the genes for an elephant do not have to tell it to obey the laws of gravity if it falls off a cliff. This means that you cannot understand gastrulation in purely genetic terms. You must build in the mathematics of buckling, too, or more precisely, you must build in whatever analog of buckling applies to the material of a blastula" (Stewart 1998).

Self-organization or self-assembly of biological structures has been called "Biological Origami" (Pivar 2004; Mahadevan 2005). It is the growing area for developmental biologists, those interested in pattern formation, that is, spatial distributions of interacting chemical and mechanical forces. It has been defined as a "buckling sphere or deformations in a growing sheet at bifurcation points followed by global depressions, wherein a growing mass of cells collectively buckles and deforms" (Mahadevan 2005). "It is an approach that assumes that each biological form is the solution to a difficult calculus problem (Cho 2004). In 1990, the geneticist Jonathan Bard wrote, "I … assert that the process of tissue formation is in many ways the cellular equivalent of molecular self-assembly and that the appropriate language in which to analyze morphogenesis is that of the differential equation" (Bard 1990; Cowin, 2004).

Many authors now demonstrate mathematical models of growth and differentiation (Cowin 2003; Gordon 2003; Mahadevan 2005). Green developed the buckling plate pattern formation, spiral phyllotaxis, for leaf and petal formation in plants (Green 1996) Developing calculations on the plant apical meristem on an elastic foundation subjected to compressive forces, Green was able to show that calculated stresses on the shell produced the same buckling pattern as the spiral phyllotaxis so often seen in plants (Green 1996; 1999; Cowin 2004). Even angles on the webs of leaf veins behave as if each were pulling on the intersection with a force proportional to its radius, just what one would expect if the veins were His' rubber tubes.

Using a model called "tensegrity," Ingber and others in his group suggest that altering the balance of physical forces can change the cells' cytoskeleton. Changing cytoskeletal geometry and mechanics can affect biochemical reactions and alter the genes that are activated and thus the proteins that are made. By simply modifying the shape of the cell, they could switch cells between genetic programs. Cells that spread were more likely to divide whereas round cells that were prevented from spreading activated a death program known as apoptosis (Chen 1997).

Jeffrey Schwartz reasons that the major reason the same two kinds of cells can produce such different structures as hair and feathers is nothing more than the structural process of invagination or evagination. "If a growing mass of cells invaginate, the result will be hair in a follicle or a skin gland. If the group of cells evaginate, the result will be scales or feathers. All of these structures occur because of the protein keratin, a protein common throughout the animal kingdom; the difference of invagination or evagination being due to the mechanical and elastic properties of the cells and the uptake of water" (Schwartz 1999). The key to this hydration, as proposed by Oster and Alberch (1982), is the accumulation of large macromolecules, named acidic mucopoly-saccharides, a key molecule in living forms and one that attracts water molecules.

Lakshminarayanan Mahadevan and Sergio Rica show a mathematical analysis of elasticity and folding where the form follows the patterns found in fluid convection and liquid crystals. They found a self-organized origami, a paper Miura-ori pattern, showing the periodic mountain-valley folds of the plant, a Hornbeam leaf. One single row of kinks along the midrib allows a folded leaf to be deployed once the bud opens. These same zigzag Miura-ori patterns are found in the drying slab of gelatin with a thin skin that forms naturally (Mahadevan 2005). Heegaard and his colleagues described a mathematical model for joint morphogeneis (Heegaard 1999); Taber and Perucchio (2000) give mathematical models for embryonic heart development in the chicken as it transforms from a single tube into a four-chambered heart (Cowin 2004). Wolfram, however, argued to change the methods from calculus-based computations for form to computational algorithm-bases. Equations were replaced by the simple step-by-step procedures of computer programs (Wolfram 2002; Cowin 2004).

GENERIC OR GENETIC?

In Alessandro Minelli's text, *The Development of Animal Form* (2003). Minelli identifies in a chapter, "Generic or Genetic," many of the problems in searching for form in developmental gene research. He outlines questions like those of Gabriel Dover (2000) who wondered "why 173 pairs of legs on a centipede? Did the centipede arrive at 173 pairs of legs by mutation and natural selection? Are 173 pairs of legs an adaptation to a special kind of environment? Or maybe it is 15 pairs of legs that have adapted to another environment? Do 16 pairs of legs give higher survival levels than 15? And how many accidents of birth are necessary to get 173 pairs of legs?" Dover also went so far as to say: "There is a naiveté about genetic determinism in both evolution and development that signifies intellectual laziness at best and shameless ignorance at worst when confronted with issues of massive complexity."

"The standard method for showing that a gene is important in, say, the development of wings in an insect, is to find a mutation of the gene that prevents wings from being formed or, even more interesting, that results in the formation of extra wings. The use of drastic gene mutations as the primary tool of investigation is a form of reinforcing practices that further convinces the biologist that any variation that is observed among organisms must be the result of genetic differences. This reinforcement then carries over into biological theory in general.... The trouble with the general scheme of explanation contained in the metaphor of development is that it is bad biology. If we had the complete DNA sequence of an organism and unlimited computational power, we could not compute the organism, because the organism does not compute itself from its genes. Any computer that did as poor a job of computation as an organism does from its genetic "program" would be immediately thrown into the trash and its manufacturer would be sued by the purchaser" (Lewontin 2000).

"There is no one-to-one relation between phenotypes and genes, yet a gene mutation is often visualized as the originator of a new phenotypic trait.... Proximate mechanisms are excluded from ultimate (evolutionary) explanations, yet a proximate mechanism (development) produces the variation that is screened by selection ... The conceptual gap that should be filled by development has been filled instead with metaphors, such as genetic programming, blueprints for organisms, and gene-environment interaction" (West-Eberhard 2003).

Minelli and others like Newman and Comper (1990) believe that genetic actions simply complement generic actions of tissues. They define generic actions as those mechanisms and physical processes that are broadly applicable to living and non-living systems. These include adhesion, surface tension, gravitational effects, viscosity, phase separation, coupling and phase-diffusion (Galbraith 1998; Salazar-Ciudad 2003). All of these may generate morphogenetic arrangements of cytoplasmic tissue and extracellular matrix components. Similarly, these authors view major types of gastrulation including epiboly, involution, and delamination as a result of physical forces. These physical forces obey the laws of physics and chemistry. For instance, Wilson, Cretekos and Helde (1995) suggest that the process of epiboly may simply be a passive, generic response to the shape and changes in the blastoderm prior to gastrulation.

But if many actions of tissues are generic, the question must be asked, "Are there developmental genes for form? Can a single change in a gene create a leg for an antenna in a fly?" Minelli (2003) answers "'Yes' in the sense that patterns of expressions of many genes are strictly limited to and correlated with specific times and events of development; 'Yes,' in so far as mutations in these genes may critically and conspicuously alter the normal course of development; but 'No' if we take any of them as directly responsible for the origin of an organ or the shaping of the body." Genes do encode transcription facts and components of intercellular signaling but these genes are much the same in animals with very different body plans (Watson 1991; Davidson; 2001). Minelli guesses that these genes have been conserved because of their generic ability to stabilize form, not because of any specific role in generating a particular body plan; for instance, genes may help in stabilizing or maintaining a given spatial arrangement. He quotes Budd (1999) that, "It is form that captures genes more than it is genes that created form."

It was Mendel who showed that superficial traits are inherited according to a choice presented by gene selection of one of two versions of the character, such as giant or dwarf, pigmented or not-pigmented, hard or soft, flexible or stiff (Mendel 1865). These are invariably simple on or off decisions between the more developed or the less developed version of the organ or organism, as though each is available in large or small size, in white or color. Even the gene jumping work of Barbara McClintock results in on and off decisions of a given gene, be it a regulatory, promoter or structural gene (Botstein 1992). Given the vertebrate phenotypic body form, uniform throughout the entire subphylum, a whale or a hummingbird will result depending on an encoded program of on or off decisions. Large hind leg genes make for rabbits and kangaroos. The entire organismal form of the animal comes in two versions, male or female, all decided by a single set of genes on a chromosome. This Mendelian inheritance has been challenged several times over the 150 years since Mendel first put forth his theory on genetic inheritance. First, by Barbara McClintock when she won the Nobel Prize in 1983 for jumping genes, by Gabriel Dover and his work with gene insertion (Dover 2000) and by the recent work of Robert Pruitt and Susan Lolle and their colleagues at Purdue University who showed the insertion of genetic material in the *Arabidopsis* (2005). Plants somehow overwrite a genetic code to get symmetry in form, replacing abnormal genes with correct copies (Lolle 2005).

Dover reminds the reader that it was William Bateson in *Materials for the Study of Variation* (1894) who first gave remarkable descriptions of insects and crustacea in which a leg would be poking out of a fly's head instead of an antenna or an antenna in place of a crab's eye. "Bateson called this type of variation 'homeosis' from which we derive the word 'homeobox'" (Dover 2000). Today these natural variations are easily produced by changes in the homeobox region, the Hox genes. Legs, antennae, wings and arms come in all sizes and shapes as first decreed by Saint-Hilaire.

In 1984, the homeobox genes were discovered that worked in the development of the fruit fly, *Drosophila*. The homeobox genes were later defined as "master control genes" and their specific role in development first described by Edward Lewis (1992). "A master control gene would be responsible for a major switch in the expression of a large number of downstream genes" (Minelli 2003). Later, it was shown that these same homeobox clusters could be found in examined multicellular organisms, not in the same number, but in the same order on the chromosome (Dover 2000). They eventually were found in chordates, echinoderms, arthropods and nematodes. When Edward Lewis, Christiane Nusslein-Volhard and Eric Wieschaus won the Nobel Prize for the identification of the genes that control segmentation in fruit flies, gene research became the central focus of developmental biology. Geneticists believed that genes were switches that instigated a cascade of other activity leading to the formation of segments and form in the egg and the embryo. A gene-centered view of development dominated the biological literature for the next few decades. What biologist has not been impressed by changes in the Hox genes, like those of the work of Duncan (1998), that cause the distal part of a *Drosophila* antenna to become transformed into a leg?

However in Hox gene research, the results are sometimes confusing. In *Drosophila*, the wing arises from the larval imaginal wing disc. The imaginal disc is a two-sided sac comprising a columnar cell layer, which can give rise to an eye, an antenna, a wing or a leg (Tabata 2004). Looking at homeotic gene regulation of these events, Carroll, in 1995, found that insect wing formation, although subject to control by Hox genes, was not created by any of them. Wing production can be extended to most trunk segments, but can be repressed in different body segments by still different Hox genes! By simple means, the switching of an on off switch, the regulatory genes, TBx4 and TBx5, legs can switch to wings and back again "Getting a leg in a particular segment of a particular species depends on patterns of 'on' and 'off' switches. Such patterns of switches differ significantly between species" (Dover 2000). Casares and Mann (1998) said that the ground state of the ventral appendages in *Drosophila* (antennae, legs, genitalia and analia) is a leg-like appendage consisting of a proximal segment and a distal tarsus. For instance the expression of two genes, *proboscipedia* and *sex combs reduced* is required for the specification of sucking mouthparts. In the absence of the activities of these two genes, an antenna forms in place of the proboscis (Percival-Smith 1997). In the absence of the activity of the gene *spineless* the distal part of a *Drosophila* antenna is transformed into a leg (Duncan 1998). The gene *distal-less* is involved in appendages but creates the eyespots on butterflies (Dover 2000). "Rather than acting as direct controllers of developmental processes, the genes act as selectors, steering development among alternative options" (Nijhout 1990).

Like Phillip Ball, Nijhout concluded that development had to be considered not just in light of gene expression, but also as a consequence of structural, chemical and physiochemical components. Nijhout also rejected the notion that the genome contains any developmental program. He uses the example of the *bicoid* gene, again in *Drosophila*, whose expression is now deemed essential in the determination of the anterior-posterior axis in the body plan. Nijhout says that what is being overlooked is the fact that the correct gradient of the *bicoid* product is established only if other products in the insect embryo are produced at the same time, allowing structural interactions in the cytoplasm. This all important bicoid gene has not been found in beetles and other arthropods.

"The role of Hox genes in determining major changes in body architecture has been extensively debated" (Minelli 2003). Minelli points out that Hox genes do not play any role in patterning the embryos or larvae of many bilaterians, such as sea urchins and polychaete annelids. In *Caenorhabditis elegans*, an essentially complete embryo is produced in the absence of any Hox gene expression (Salser 1994; Van Auken 2000). And in many developing organisms, Hox genes contribute only to the adult form and not to the larval forms (Williamson 2003; Peterson 2000).

According to Evelyn Keller (2000) in *The Century of the Gene*, the notion of a genetic program depends on the common mistake of equating the distinction between genetic and epigenetic with the distinction between program and data. The genetic information present in a zygote is usually thought of as a computer program, but it might be more accurate to consider it as a set of data that needs to be processed by some internal program.

For instance, the *eyeless* (*ey* gene) in *Drosophila*, a homologue of the *Pax6* in *Xenopus* was long thought to be a master control gene for eye morphogenesis (Halder 1995; Chow 1999). The involvement of *Pax6* homologues was found in squid (Tomarev 1997), amphioxus (Glardon 1998), nemerteans (Kmita-Cunisse 1998) and planarians (Pineda 2000). Developmental geneticists were quick to conclude that the origin of the eye is monophyletic throughout the animal kingdom (Gehring 1999, 2000). However, later, Wagner (2000) found that the *Pax6* gene initiated the development of needed proteins for the early development of light sensitive epithelia, a simple precursor of both the camera eye of vertebrates and squids and the compound eye of arthropods. The visual sensory apparatus in all these cases is epithelial and derived from ectoderm. In fact *Pax6* expression in vertebrates is not limited to the eye, but extends to nasal placodes, parts of the brain and spinal cord (Li 1994; Amirthalingam 1995). In squid it is even expressed in the arms, and in sea urchins, in the tube feet (Czerny 1995). *Sonic hedgehog* is active in limb bud signaling and can grow extra digits when transplanted onto a budding limb. However, the same *Sonic hedgehog* protein, acting as a growth factor, signals the development of the brain and spinal cord (Riddle 1999). Minelli (2003) asks, "Is modern molecular genetic evidence more convincing than this?"

The very existence of master control genes has recently been called into question. Of the seven *Drosophila* genes that were originally considered "Master Control Genes" for the eye, none have passed the criterion of producing homeosis, that is, gene expression when the gene is transferred from one site to another. Davidson (2001) expresses serious doubt about the concept of master control genes and calls them a "fantasy of earlier days." "The role of genes in morphogenesis is likely to always be indirect" (Minelli 2003). Shubin and Marshall (2000) go even further, believing that the specific role of the "so-called developmental genes" is rather mundane. "Most of them regulate the rate of cell proliferation or determine cell adhesion properties or act as transcription factor, binding to specific DNA segments. The scientific literature of the last decade is full of examples of genes whose temporally and spatially correct expression is critical for the production of complex, specific features, but at the time are known to encode for proteins whose mechanism is quite unspecific" (Schejter 1993).

Ellen Larsen, a prominent biologist in the genetics of *Drosophila*, similarly believes that relatively few changes in cell behavior may be responsible for large morphological changes without unraveling the rest of development (Larsen 2003).

Davidson, in *Genomic Regulatory Systems: Development and Evolution* (2001), also writes that there are no genes specific to a given body plan. He says, "There are no insect genes, no sea urchin genes, and no vertebrate genes. Furthermore, differences between a beetle and a frog are not a matter of building stones, nor are they a matter of the kind of paper and ink used to draw a blueprint of their respective body plans." Davidson is not alone in this view. Minelli (2003) pointed out that Wagner, Chiu and Laubichler (2000) conclude much the same thing, "Evidence at hand does not demonstrate that any particular gene is instrumental in the origin of major characters, such as insect wings." One gene can seem to control the development of many forms. For instance, two homeobox genes, *nubbin* and *apterous* are expressed in similar ways both in the wings of a fruit fly and in the limbs of *Artemia*, a shrimp-like crustacean (Schwartz 1999). Sometimes something as simple as the stimulation of cell growth is enough to produce changes in body form and pattern.

Many call the current interest in developmental genetics "default morphology," meaning that to attribute formation of a given body part to the specific activity of one or a few genes is unwarranted. "It is incorrect to conclude that a gene must be a major or pivotal one, if a particular process does not work in its absence" (Lawrence 2001). "The gene is ignorant of its role in development" (Dover 2000). "Developmental genetics is discovering more and more examples of default morphology manifested when some gene activity is lacking. One could ideally push gene silencing further and further, until a minimum gene number is found, by which a given level of morphological complexity is still generated or, better, saved and transmitted" (Minelli 2003). Some of the work in Hox genes today can be compared to Spemann and Mangold's organizer (Spemann 1924). By surgically transplanting the lip of the developing gastrula of salamanders, they obtained two mirror image embryonic forms for which Hans Spemann later won the Nobel Prize. However, later it was found that a dead organizer worked just as well (Twitty 1966) and in fact everything is an inducer, including pH changes, sodium chloride, lithium chloride, basic proteins from mouse kidney, protein from guinea pig, extract of nine-day chick embryos, worm fragments, any organ from fishes, amphibia, birds, mammals and even snail foot muscles. Richard Gordon in a seminar given at the Konrad Lorenz Institute in Austria (2002) lists the current "in fashion" inducers as the homeobox gene *goosecoid*, fibroblast growth factor, *activan* and *noggin*.

MECHANICAL FORCE

"I see no necessity for the belief that the eye was expressly designed, without however concluding that everything, especially Man is the result of 'brute force'" (Darwin, letter to Asa Gray, 1860). "Mechanics is shown to be an important, perhaps, central component to the differentiation and development of embryos" (Gordon 2003). Are there forces other than gene activity that are responsible for cell division and form? Wilhelm His explained mechanical force on development using the analogy of the rubber tube to explain neural tube development and the buckling of laminates. "To think that heredity will build organic beings without mechanical means is a piece of unscientific mysticism ... Embryology and morphology cannot proceed independently of all reference to the general laws of matter, to the laws of physics and mechanics. This proposition would, perhaps, seem indisputable to every natural philosopher; but, in morphological schools, there are very few who are disposed to adopt it with all its consequences" (His 1888). His's mechanical approach to biology had almost vanished from contemporary literature with the onset of modern genetics (Gordon 2003). D'Arcy Thompson's contribution in 1917, *On Growth and Form*, through extremely notable, had been archived to the back shelves of contemporary libraries. It is only with the current interest in mathematics, computer science, physics, and their relation to form have the ideas of Wilhelm His, Jacques Loeb, and Johann Wolfgang Goethe again taken center stage (His 1888; Loeb, 1912; Thompson 1917).

"The natural language of pattern and form is mathematics.... The main point is that mathematics enables us to get to grips with the *essence* of pattern and form – to describe it at its most fundamental level, and thereby to see most clearly what features need to be reproduced by an explanation or a model.... The real advantage of using mathematics to describe form is that it makes the problem *algorithmic*. An algorithm is a sequence of logical steps that a computer program, say, must execute to carry out a certain task. Complex shapes like those seen here are often most easily described not in terms of 'what goes where' but by an algorithm that generates them. Once we know the mathematical algorithm, we can start to ask what kind of physical processes might provide a form-generating rule to which the algorithm is a good approximation. In this way, the mystery of form and pattern becomes much more clearly defined" (Ball 1999).

Some actions of cells and tissue to mechanical force are accepted as natural phenomena. For instance, it has been known for a long time that cells behave differently when placed on different surfaces, becoming more spherical or more elongated as the substrate changes (Harris 1980). Distortive mechanical pressure on many cell types can induce cell division. For instance, genes that strengthen the cells lining blood vessels can be activated by the stretching effect of high blood pressure. This mechanical stretching expresses the TIGR/MYOC and Oculomedin gene and subsequent protein expression (Tamm 1996). The TIGR mRNA/protein expression is essential to glaucoma pathogenesis (Nguyen 1998). This action is thought to occur by some interaction with the plasma membrane that senses a tension in the cell membrane. It was suggested that the mechanosensors within the membrane are stretch-activated, by sensing the mechanical stretch of the membrane (Hamill 1992; 2001; Sackin 1995). Another candidate for mechanosensors is integrins, transmembrane receptors that couple components of the extracellular matrix with the actin of the cytoskeleton, lying directly under and attached to the membrane (Wang 1993). The cytoskeleton may form a complex with membrane proteins such as ion channels, regulating their responsiveness to external forces (Watson 1991).

"Mechanical stress is one of the major epigenetic factors regulating differentiation of skeletal tissues including chondrocytes, osteoblasts and fibroblasts" (Buckwalter 1995; Takahashi 2005). The mechanical loading of long bones induces immediate early gene expression in genes such as Erg-1 that control cell division and proliferation (Dolce 1996). Bone tissue can adjust its mass and architecture to functionally adapt to the mechanical environment (Tang 2004) and the response of the osteoblasts is dependent on the magnitude of the strain. Thus cells adjust their response to adapt to the changing environmental stimulation (Tang 2004). Gene induction by mechanical forces is found also in cardiac cells. In cultured neonatal cardiac myocytes, mechanical stretch causes the activation of the synthesis of a number of proteins (for review see Hamill 2001).

Emmanuel Farge, a physicist working at the Curie Institute in Paris, challenges the idea that genetic programs are necessary to twist, bend and bud embryonic tissue into heads, limbs and internal organs. He looked at the idea that physical forces can guide and shape the earliest stages of life, the developing embryo. *In Current Biology* (2003), reviewed in *Nature* (2003), he reports that gene expression can be modulated by the mechanical strains of morphogenetic movements, which can be as simple as growth within a confined membrane. When Farge briefly flattened three-hour old fruit fly embryos, their surface switched on a gene called "*Twist*" which is normally expressed only on the ventral side. Within eight minutes of compression, *Twist* expression had spread to encompass the entire embryo, even in embryos in which the dorsal-ventral patterning system had been eliminated (Scott 2003).

Farge proposed that the compression of the embryo was inducing β-catenin (the fly homolog is Armadillo) to move into the nucleus (Farge 2003). β-catenin is a multifunctional protein that plays an essential role in cell-cell adhesion and nuclear gene activating signaling (Sanson 1996). Once in the nucleus, Armadillo mediates the expression of *Twist* and *Snail* developmental genes.

Farge showed that the mechanical stress of early *Drosophila* embryos, that is, a unilateral 10% deformation for five minutes, caused the translocation of Armadillo from the plasma membrane to the nucleus and the expression of the *Twist* gene. When the embryonic cells of the *Drosophila* were compressed during the first phase of germ band *Twist* expression increased eightfold.

The *Twist* gene was not expressed in mutants that blocked morphogenetic movement of the germ-band extension. However, when Farge rescued the mutants with gentle compression, the *Twist* gene was expressed again. Mechanical force was necessary for gene expression and cell division. In a growing cell mass during early blastula development, cell growth is the necessary force for division. "These cells in the embryonic *Drosophila* are normally subjected to pressure that is made by the extending germ band at the posterior pole and the pressure is transmitted to the anterior pole, compressing the cells against the invagination of the mesoderm and foregut" (Heinrichs 2003). Instead of genes controlling morphogenic movements, morphogenic movements were controlling gene activity.

"You can reprogram the embryo mechanically at the beginning of its life," concludes Farge. Farge further speculates, "that mechanical induction may be an ancient mechanism for inducing gut formation. This could have evolved from a primitive reflex response to mechanical deformations. Such a response might have been the phagocytosis of particles in response to physical contact which has been proposed to be the 'feeding-response' of the earliest organisms."

"Morphogenetic cell migrations or changes in the embryonic topology, e.g. at gastrulation, are already known to play an important role through classical induction. These movements cause changes in the cell environment, enabling gene induction via new local interaction" (Brouzés 2004).

The work of Farge's lab looked specifically at mesoderm invagination, the first morphogenetic movement of gastrulation and ventral furrow invagination in the early *Drosophila* embryo. This cell layer folding has been correlated with the trapezoidal cell shape change initiated by apical constriction, one of the first mechanical movements that begin to shape the embryo. This coordination of apical constriction is essential for proper mesoderm invagination (Keller 2003). Following this first step of cell shape changes, the cells progressively lose their adhesive properties, recover mitotic activity and migrate to colonize the ectoderm. This transition occurs simultaneously with cell differentiation (Leptin 1999).

During embryonic development, mechanical action is particularly important in that the extracellular matrix for the formation of bone can only arise as a result of mechanical stress. Substantial stress is generated during embryonic growth by cell crowding. The pressure of simple growth and division within a confined membrane generates a physical force, which may then control early embryonic development.

This same induction of gene activity by mechanical pressure may also play a role in the induction of genes in segmentation. The observed patterns of segmentation may be accounted for as stress patterns imposed by the bending of the embryonic axis. Something as simple as uneven growth rates producing an embryonic bend with bands of cells can be the generating force for gene induction (Huxley 1924). Tissue tension in explants of gastrula *Xenopus* embryos is necessary for the development of a normal set of rudiments and for differentiation (Brouzés 2004; Beloussov 1988) and mechanical contraction waves were reported to correlate with primary neural induction during development in azolotyl embryos (Brouzés 2004; Brodland 1994).

NATURAL SELECTION

If genes do not determine form, how can natural selection acting on mutations within the genes guide evolution? Jeffrey Schwartz, in his book *Sudden Origins* (1999), discusses the phenomenon as related to genetic homeostasis. "Natural selection acts to maintain stability." This concept was first expressed by Williams (1966) in *Adaptation and Natural Selection: A Critique of Some Current Evolutionary Thought.* He wrote that natural selection and mutation could not alone explain the origin of a species. It was his opinion that mutation interrupts the stability that selection tries to maintain. Mutation alone may not account for all the species variations. "Point mutations, which seem to occur most frequently in the third base of a triplet codon in the DNA sequence encoding a specific molecule, tend to have no expressed effect on the molecule's structure and its function in the organism. Consequently a point mutation is essentially silent" (Schwartz 1999). Gould and Eldredge (1977) also argued the inconsequential effect of point mutations: "We do not see how point mutations in structural genes can lead, even by gradual accumulation, to new morphological designs … The near identity of humans and chimps for structural genes, and the evidence of major regulatory change indicated by human neoteny provides an important confirmation.

"As an explanation for natural form, natural selection is not entirely satisfying. Not because it is wrong, but because it says nothing about mechanism.… Once you start to ask the 'how?' of mechanism, you are up against the rules of chemistry, physics, and mechanics, and the question becomes not just 'is the form successful?' but 'is it physically possible?'"

According to Gabriel Dover, natural selection is nothing more than the one-off, passive outcome of a unique set of interactions of newly established, unique individuals with the local environment, in each generation "There is no force directly selecting phenotypes or selecting 'selfish genes.'" Speaking of the question of the centipede's legs, Dover goes on to say, "Did all the legs on a centipede arise only through the inevitable provision, by natural selection, of novel solutions to external problems: Or could they also have arisen and spread by a molecular drive process in a small population of proto-centipedes, who, without too much fuss, got on with the business of living, reproducing and generally exploiting, with their newly acquired legs, some parts of their existing environment that were previously inaccessible: In short, did the centipede need to evolve all its legs?" (Dover 2000).

In a paper in *International Journal of Developmental Biology* (2003), Richard Gordon and Cameron Melvin write that "… only when principles borrowed from mathematics, fluid mechanics, materials science, etc. are applied to classical problems in developmental biology, will sufficient comprehension be achieved to permit successful understanding and therapeutic manipulation of embryos. As it now stands, biologists seldom possess either skills or interest in those areas of endeavor. Thus we have determined that it is easier to educate engineers in the principles of developmental biology than to help biologists deal with the complexities of engineering."

It was "D'Arcy Thompson's thesis that biology cannot afford to neglect physics, in particular that branch of it that deals with the mechanics of matter.... For him this did not answer questions about causes; it merely relocated the question. A physicist, on the other hand, 'finds "causes" in what he has learned to recognize as fundamental properties ... or unchanging laws, of matter and energy....'

There are, furthermore, physical properties that biological structures possess, such as surface tension, electrical charge and viscosity. These are all relevant to the way that cells work, but gene-hunting cannot tell us much at all about what their role is" (Ball 1999).

References

Amirthalingam, K., J.B. Lorens, B.O. Saetre, E. Salaneck, A. Fjose. 1995. Embryonic expression and DNA-binding properties of zebrafish *Pax-6*. *Biochemical and Biophysical Research Communications* 215(1): 122–128.

Ast, G. 2005. The alternative genome. *Scientific American* 292(4): 58–65.

Ball, P. 1999. *The Self-Made Tapestry: Pattern Formation in Nature*. Oxford: Oxford University Press.

Bard, J. 1990. *Morphogenesis*. Cambridge: Cambridge University Press.

Bateson, W. 1894. *Materials for the Study of Variation, treated with especial regard to discontinuity in the origin of species*. London: Macmillan and Co.

Beloussov, L.V., A.V. Lakirev, I.I. Naumidi. 1988. The role of external tensions in differentiation of *Xenopus laevis* embryonic tissues. *Cell Differ. Dev.* 25: 165–176.

Bonner, J. 1993. *Life Cycles*. Princeton: Princeton University Press.

Botstein, D., N. Fedoroff, eds. 1992. *The Dynamic Genome: Barbara McClintock's Ideas in the Century of Genetics*. Cold Spring Harbor: Cold Spring Harbor Laboratory Press.

Brenner, S., J.H. Miller, eds. 2001. *Encyclopedia of Genetics*. San Diego: Academic Press.

Brodland, G.W., R. Gordon, M.J. Scott, N.K. Björklund, K.B. Luchka, C.C. Martin, C. Matuga, M. Globus, S. Vethamany-Globus, D. Shu. 1994. Furrowing surface contraction wave coincident with primary neural induction in amphibian embryos. *J. Morphology* 219: 131–142.

Brouzés, E., E. Farge. 2004. Interplay of mechanical deformation and patterned gene expression in developing embryos. *Current Opinion in Genetics & Development* 14: 367–374.

______, W. Supatto, E. Farge. 2005. Is mechano-sensitive expression of *Twist* involved in mesoderm formation? *Biology of the Cell* 96(7): 471–477.

Buckwalter, J.A., M.J. Glimcher, R.R. Cooper, R. Recker. 1995. Bone biology II: formation, form, modeling, remodeling and regulation of cell function. *J. Bone Joint Surg. Am* 77: 1276–1289.

Budd, G.E. 1999. Does evolution in body patterning genes drive morphological change—or vice versa? *BioEssays* 21: 326–332.

Carroll, S. B., S.D. Weatherbee, J.A. Langeland. 1995. Homeotic genes and the regulation and evolution of insect wing number. *Nature* 375: 58–61.

Casares, F., R.S. Mann. 1998. Control of antenna versus leg development in *Drosophila*. *Nature* 392: 723–726.

Chen, C.S., M. Mrksich, S. Huang, G.M. Whitesides, D.E. Ingber. 1997. Geometric control of cell life and death. *Science* 276(5317): 1425–1428.

Chen, T., T. Hiroko, A. Chaudhuri, F. Inose, M. Lord, S. Tanaka, J. Chant, A. Fujita. 2000. Multigenerational cortical inheritance of the Rax2 protein in orienting polarity and division in yeast. *Science* 290: 1975–1978.

Cho, A. 2004. Life's patterns, no need to spell it out? *Science* 303(5659): 782–783.

Chow, R. L., C.R. Altmann, R.A. Lang, A. Hemmati-Brivanlou. 1999. *Pax6* induces ectopic eyes in a vertebrate. *Development* 126: 4213–4222.

Cowin, S.C. 2000. How is a tissue built? *Journal Biomechanical Engineering* 122: 553–569.

______. 2004. Tissue growth and remodeling. *Annu. Rev. Biomed. Eng.* 6: 77–107.

Czerny, T., M. Busslinger. 1995. DNA-binding and transactivation properties of *Pax-6*: Three amino acids in the paired domain are responsible for the different sequence recognition of *Pax-6* and BSAP (*Pax-5). Mol. Cell. Biol.* 15(5): 2858–2871.

Darwin, C. 1860. Letter to American biologist Asa Gray.

Davidson, E. 2001. *Genomic Regulatory Systems: development and evolution.* San Diego: Academic Press.

Dolce, C., A.J. Kinniburgh, R. Dziac. 1996. Immediate early-gene induction in rat osteoblastic cells after mechanical deformation. *Arch. Oral Biol.* 41: 1101–1108.

Dover, G. 2000. *Dear Mr. Darwin: Letters on the Evolution of Life and Human Nature.* London: Phoenix Press.

Duncan, D.M., E.A. Burgess, I. Duncan. 1998. Control of distal antennal identity and tarsal development in *Drosophila* by spineless-aristapedia, a homolog of the mammalian dioxin receptor. *Genes and Development* 12(9): 1290–1303.

Edelman, G.M. 1988. *Topobiology: An Introduction to Molecular Embryology*. New York: Basic Books.

Farge, E. 2003. Mechanical induction of *Twist* in the *Drosophila* foregut/stomodeal primordium. *Curr. Biol.* 13: 1365–1377.

Farge, M. 1992. Wavelet transforms and their applications to turbulence. *Annual Review of Fluid Mechanics* 24: 395–457.

Fusco, G., C. Brena, A. Minelli. 2000. Cellular processes in the growth of lithobiomorph centipedes (Chilopoda Lithobiomorpha). A cuticular view. *Zoologischer Anzeiger* 239: 91–102.

Galbraith, C.G, M.P. Sheetz. 1998. Forces on adhesive contacts affect cell function. *Curr. Opin. Cell Biol.* 10(5): 566–571.

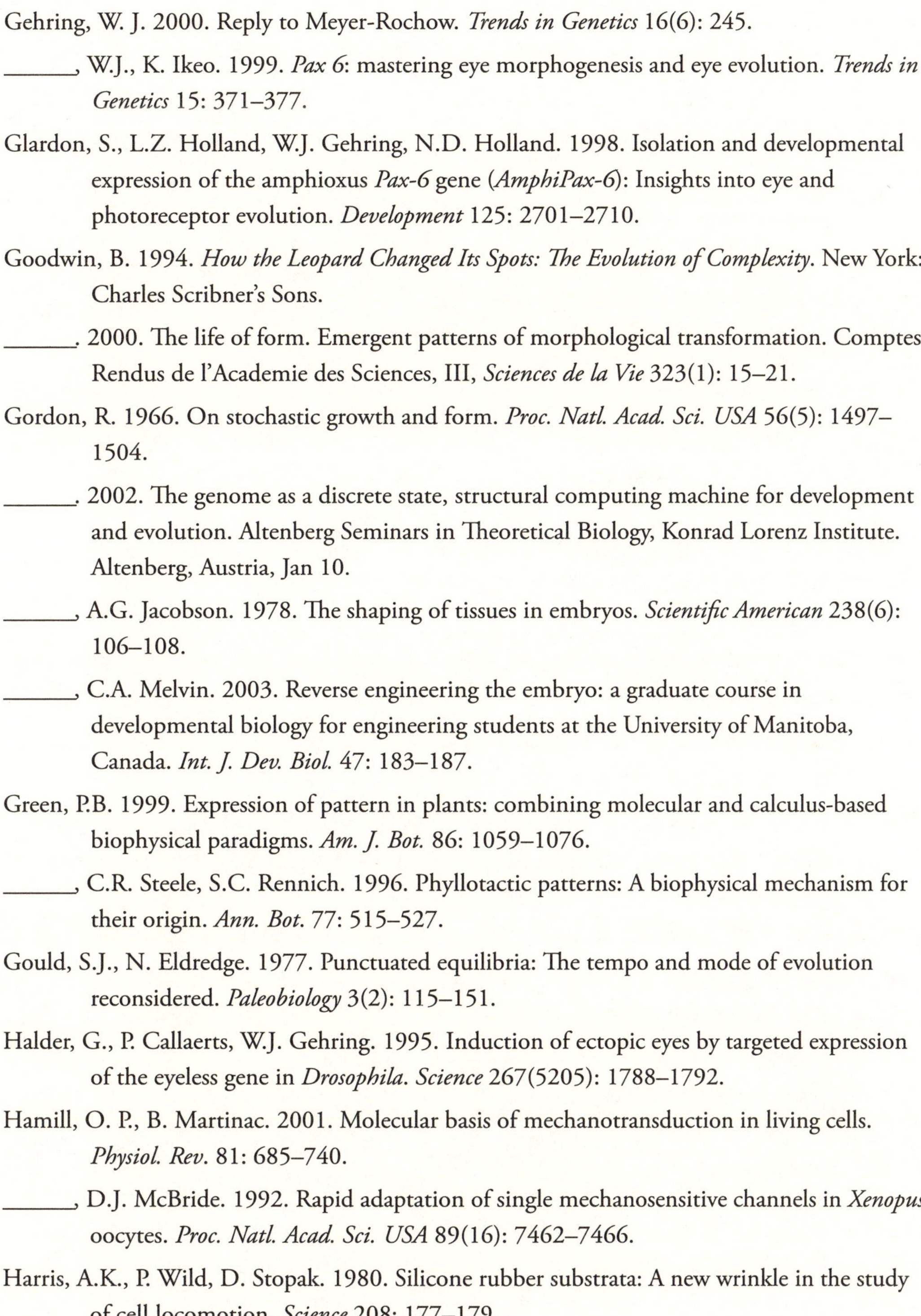

Gehring, W. J. 2000. Reply to Meyer-Rochow. *Trends in Genetics* 16(6): 245.

______, W.J., K. Ikeo. 1999. *Pax 6*: mastering eye morphogenesis and eye evolution. *Trends in Genetics* 15: 371–377.

Glardon, S., L.Z. Holland, W.J. Gehring, N.D. Holland. 1998. Isolation and developmental expression of the amphioxus *Pax-6* gene (*AmphiPax-6*): Insights into eye and photoreceptor evolution. *Development* 125: 2701–2710.

Goodwin, B. 1994. *How the Leopard Changed Its Spots: The Evolution of Complexity*. New York: Charles Scribner's Sons.

______. 2000. The life of form. Emergent patterns of morphological transformation. Comptes Rendus de l'Academie des Sciences, III, *Sciences de la Vie* 323(1): 15–21.

Gordon, R. 1966. On stochastic growth and form. *Proc. Natl. Acad. Sci. USA* 56(5): 1497–1504.

______. 2002. The genome as a discrete state, structural computing machine for development and evolution. Altenberg Seminars in Theoretical Biology, Konrad Lorenz Institute. Altenberg, Austria, Jan 10.

______, A.G. Jacobson. 1978. The shaping of tissues in embryos. *Scientific American* 238(6): 106–108.

______, C.A. Melvin. 2003. Reverse engineering the embryo: a graduate course in developmental biology for engineering students at the University of Manitoba, Canada. *Int. J. Dev. Biol.* 47: 183–187.

Green, P.B. 1999. Expression of pattern in plants: combining molecular and calculus-based biophysical paradigms. *Am. J. Bot.* 86: 1059–1076.

______, C.R. Steele, S.C. Rennich. 1996. Phyllotactic patterns: A biophysical mechanism for their origin. *Ann. Bot.* 77: 515–527.

Gould, S.J., N. Eldredge. 1977. Punctuated equilibria: The tempo and mode of evolution reconsidered. *Paleobiology* 3(2): 115–151.

Halder, G., P. Callaerts, W.J. Gehring. 1995. Induction of ectopic eyes by targeted expression of the eyeless gene in *Drosophila*. *Science* 267(5205): 1788–1792.

Hamill, O. P., B. Martinac. 2001. Molecular basis of mechanotransduction in living cells. *Physiol. Rev.* 81: 685–740.

______, D.J. McBride. 1992. Rapid adaptation of single mechanosensitive channels in *Xenopus* oocytes. *Proc. Natl. Acad. Sci. USA* 89(16): 7462–7466.

Harris, A.K., P. Wild, D. Stopak. 1980. Silicone rubber substrata: A new wrinkle in the study of cell locomotion. *Science* 208: 177–179.

Heegaard, J.H., G.S. Beaupré, D.R. Carter. 1999. Mechanically modulated cartilage growth may regulate joint surface morphogenesis. *J. Orthop. Res.* 17(4): 509–517.

Heinrichs, A. 2003. Gene expression: Let's do the twist. *Nature Reviews Genetics* 4: 760.

His, W. 1888. On the principles of animal morphology. *Roy. Soc. Edinburgh Proc.* 15: 287–298.

Huxley, J. 1924. Constant differential growth ratios and their significance. *Nature* 114: 895–896.

Ingber, D.E. 1998. The architecture of life. *Scientific American* 278: 48–57.

Kauffman, S.A. 1993. *The Origins of Order: Self-Organization and Selection in Evolution.* New York: Oxford University Press.

______. 2000. *Investigations.* New York: Oxford University Press.

Keller, E.F. 2000. *The Century of the Gene.* Cambridge: Harvard University Press.

Keller, R. 2002. Shaping the vertebrate body plan by polarized embryonic cell movements. *Science* 298(5600): 1950–1954.

______, L.A. Davidson, D.R. Shook. 2003. How we are shaped: The biomechanics of gastrulation. *Differentiation* 71(3): 171–205.

Kmita-Cunisse, M., F. Loosli, J. Bierne, W.J. Gehring. 1998. Homeobox genes in the ribbonworm Lineus sanguineus: Evolutionary implications. *Proc. Natl. Acad. Sci. USA* 95(6): 3030–3035.

Le Guyader, H. 2004. *Geoffroy Saint-Hilaire, A Visionary Naturalist.* Chicago: University of Chicago Press.

Larsen, E. 2003. Genes, cell behavior, and the evolution of form. *Origination of Organismal Form*, G. B. Müller, S. A. Newman, eds., pp. 119–132. Cambridge, MA: Bradford Books/MIT Press.

Lawrence, P. A. 2001. Science or alchemy? *Nature Reviews Genetics* 2(2): 139–142.

Leptin, M. 1999. Gastrulation in *Drosophila*: The logic and the cellular mechanisms. *EMBO J.* 18(12): 3187–3192.

Lewis, E.B. 1992. Clusters of master control genes regulate the development of higher organisms. *JAMA* 267(11): 1524–1531.

Lewontin, R. 2000. *It Ain't Necessarily So: The Dream of the Human Genome and Other Illusions.* New York: New York Review of Books.

______. 2000. *The Triple Helix: Gene, Organism and Environment.* Cambridge: Harvard University Press.

Li, H.S., J.M. Yang, R.D. Jacobson, D. Pasko, O. Sundin. 1994. *Pax-6* is first expressed in a region of ectoderm anterior to the early neural plate: Implications for stepwise determination of the lens. *Dev. Biol.* 162: 181–194.

Loeb, J. 1912. *The Mechanistic Conception of Life: Biological Essays.* Chicago: The University of Chicago Press. 1964 Reprint, Cambridge: Harvard University Press.

Lolle, S.J., J.L. Victor, J.M. Young, R.E. Pruitt. 2005. Genome-wide non-mendelian inheritance of extra-genomic information in *Arabidopsis. Nature* 434: 505–509.

Mahadevan, L., S. Rica. 2005. Self-organized origami. *Science* 307(5716): 1740.

McClintock, B. 1983. Nobel Prize in Physiology or Medicine.

Mendel, G. 1865. Experiments in Plant Hybridization. *Proceedings of the of the Natural History Society of Brünn.*

Miodownik, M.A. 2003. Using Mechanics to Map Genotype to Phenotype. In *On Growth, Form and Computers*, S. Kuman and P.J. Bentley, eds., pp. 203–219. New York: Elsevier Academic Press.

Minelli, A. 2003. *The Development of Animal Form: Ontogeny, Morphology, and Evolution.* Cambridge: Cambridge University Press.

Newman, S.A., W.D. Comper. 1990. 'Generic' physical mechanisms of morphogenesis and pattern formation. *Development* 110(1): 1–18.

Müller, W.A. 1996. *Developmental Biology.* New York: Springer.

Müller, G.B., S.A. Newman. 2003. Origination of organismal form: The forgotten cause in evolutionary theory. In *Origination of Organismal Form*, G.B. Müller and S.A. Newman, eds., pp. 3–10. Cambridge: Bradford Books/MIT Press.

Nijhout, H.F. 1990. Metaphors and the role of genes in development. *BioEssays* 12: 441–446.

Nguyen, T.D., P. Chen, W.D. Huang, H. Chen, D. Johnson, J.R. Polansky. 1998. Gene structure and properties of TIGR, an olfactomedin-related glycoprotein cloned from glucocorticoid-induced trabecular meshwork cells. *J. Biol. Chem.* 273: 6341–6350.

Oster, G.F., P. Alberch. 1982. Evolution and bifurcation of developmental programs. *Evolution* 36: 444–459.

Oyama, S., 2000. *The Ontogeny of Information: Developmental Systems and Evolution* (2nd ed.). Durham, NC: Duke University Press.

Peterson, K.J., S.Q. Irvine, R.A. Cameron, E.H. Davidson. 2000. Quantitive assessment of Hox complex expression in the indirect development of the polychaete annelid *Chaetopterus* sp. *Proc. Natl. Acad. of Sci. USA* 97(9): 4487–4492.

Percival-Smith, A., J. Weber, E. Gilfoyle, P. Wilson. 1997. Genetic characterization of the role of the two Hox proteins, proboscipedia and *sex combs reduced*, in determination of adult antennal, tarsal, maxillary palp and proboscis identities in *Drosophila melanogaster*. *Dev.*124(24): 5049–62.

Pineda, D., J. Gonzales, P. Callaerts, K. Ikeo, W.J. Gehring, E. Salo. 2000. Searching for the prototypic eye genetic network: *Sine oculis* is essential for eye regeneration in planarians. *Proc. Natl. Acad. Sci. USA* 97(9): 4525–4529.

Pivar, S., 2004. *Lifecode: The Theory of Biological Self Organization.* New York: Ryland Press.

Riddle, R.D., C.J. Tabin. 1999. How limbs develop. *Scientific American* Feb: 74–79.

Ridley, M. 1999. *Genome: The Autobiography of a Species in 23 Chapters.* New York: Harper Collins.

Sackin, H. 1995. Stretch-activated ion channels. *Kidney Intl.* 48: 1134–1147.

Salazar-Ciudad, I., J. Jernvall, S. Newman. 2003. Mechanisms of pattern formation in development and evolution. *Dev.* 130: 2027–2037.

Salser, S., C. Kenyon. 1994. Patterning C. elegans: Homeotic cluster genes, cell fates and cell migrations. *Trends in Genetics* 10(5): 159–164.

Sanson, B., P. White, J. Vincent. 1996. Uncoupling cadherin-based adhesion from wingless signaling in *Drosophila. Nature* 383: 627–630.

Schejter, E.D., E. Wieschaus. 1993. Bottleneck acts as a regulator of the microfilament network governing cellularization of the *Drosophila* embryo. *Cell* 75(2): 3733–85.

Schwartz, J.H. 1999. *Sudden Origins: Fossils, Genes, and the Emergence of Species.* New York: John Wiley & Sons.

Scott, I.C., D.Y. Stainier. 2003. Developmental biology: Twisting the body into shape. *Nature* 425(6957): 461–463.

Shubin, N.H., C.R. Marshall. 2000 Fossils, genes, and the origin of novelty. *In* Deep time: *Paleobiology's* perpective, D.H. Erwin, S.L. Wing, eds. *Paleobiology* 26(Suppl. to 4): 324–340.

Sonnebom, T. 1965. Cytoplasmic inheritance of the organization of the cell cortex in Paramecium aurelia. *Proc. Natl. Acad. Sci. USA* 53: 275–282.

Spemann, H.H., H. Mangold. 1924. Über induktion von embryonanlagen durch implantation artfremder organisatoren. *Wilh. Roux Arch. EntwMech. Organ.* 100: 599–638.

Supatto, W., D. Débarre, B. Moulia, E. Brouzés, J.L. Martin, E. Farge, E. Beaurepaire. 2005. *In vivo* modulation of morphogenetic movements in *Drosophila* embryos with femtosecond laser pulses. *Proc. Natl. Acad. Sci. USA* 102(4): 1047–1052.

Stewart, I. 1998. *Life's Other Secret: The New Mathematics of the Living World.* New York: John Wiley & Sons.

Tabata, T., Y. Takei. 2004. Morphogens, their identification and regulation. *Dev.* 131: 703–712.

Taber, L.A., R. Perucchio. 2000. Modeling heart development. *J. Elasticity* 61: 165–197.

Takahashi, I., Y. Hatakeyama, F. Terao, Y. Sasano, J. Sugawara. 2005. Mechanical stretch induces ERK1/2 phosphorylation in micromass culture. Seq. 271: Effects of Mechanical Factors on Cells. Baltimore Convention Center, 12 March.

Tamm, E.R., P. Russell, D.H. Johnson, J. Piatigorsky. 1996. Human and monkey trabecular meshwork accumulate alpha B-crystallin in response to heat shock and oxidative stress. *Invest. Ophthalmol. Vis. Sci.* 37(12): 2402–2413.

Tang, L., Y. Wang, J. Pan, S. Cai. 2004. The effect of step-wise increased stretching on rat calvarial osteoblast collagen production. *J. Biomechanics* 37(1): 157–161.

Tartar, V. 1961. *The Biology of Stentor*. Oxford, UK: Pergamon Press.

Tetsuya, T., Y. Takei. 2004. Morphogens, their identification and regulation. *Dev.* 131: 703–712.

Thompson, J.M.T. 1982. *Instabilities & Catastrophes in Science and Engineering*. New York: John Wiley & Sons.

Tomarev, S.I., P. Callaerts, L. Kos, R. Zinovieva, G. Halder, W.J. Gehring, J. Piatigorsky. 1997. Squid *Pax-6* and eye development. *Proc. Natl. Acad. Sci. USA* 94: 2421–2426.

Twitty, V.C. 1966. *Of Scientists and Salamanders*. San Francisco: W.H. Freeman and Co.

Van Auken, K., D.C. Weaver, L.G. Edgar, W.B. Wood. 2000. *Caenorhabditis elegans* embryonic axial patterning requires two recently discovered posterior-group Hox genes. *Proc. Natl. Acad. Sci. USA* 97(9): 4499–4503.

Vogel, K.G., T.J. Koob. 1989. Structural specialization in tendons under compression. *Int. Rev. Cytol.* 115: 267–293.

Wagner, G.P., C-H Chiu, M. Laubichler. 2000. Developmental evolution as a mechanistic science: The inference from developmental mechanisms to evolutionary processes. *American Zoologist* 40(5): 819–831.

Wang, N., J.P. Butler, D.E. Ingber. 1993. Mechanotransduction across the cell surface and through the cytoskeleton. *Science* 260(5111): 1124–1127.

Watson, P.A. 1991. Function follows form: Generation of intracellular signals by cell deformation. *FASEB J.* 5: 2013–2019.

Weitzel, H., M. Illies, C. Byrum, X. Ronghui, A. Wikramanayake, C. Ettensohn. 2004. Differential stability of ß-catenin along the animal-vegetal axis of the sea urchin embryo mediated by dishevelled. *Dev.* 131: 2947–2956.

West-Eberhard, M.J. 2003. *Developmental Plasticity and Evolution*. New York: Oxford University Press.

Williams, G.C. 1966. *Adaptation and Natural Selection: A Critique of Some Current Evolutionary Thought*. Princeton: Princeton University Press.

Williamson, D.I. 2003. *The Origins of Larvae*. Boston: Kluwer Academic Publishers.

Wilson, E.T., C. Cretekos, K.A. Helde. 1995. Cell mixing during early epiboly in zebrafish embryo. *Developmental Genetics* 17(1): 6–15.

Wolfram, S. 2002. *A New Kind of Science*. Champaign, IL: Wolfram Media.

Zimmer, C. 2001. *Evolution: The Triumph of an Idea*. New York: Harper Collins Publishers.

Gregor Mendel, Thomas Henry Huxley, and the Acceptance of New Scientific Theories

Richard Milner

"My time will come," said the brilliant scientist-monk Gregor Mendel. Working with thousands of pea plants in the monastery garden at Brunn, Austria, he made the crucial breakthrough (1865) in understanding how heredity works. But no biologist recognized his achievement until thirty-five years later, though it was the key to evolution they had all been seeking. Why?

To begin with, Mendel published very little, only two short papers in the journal of a local natural history society. Although he sent reprints to some of the important botanists of his day, many ignored it simply because he wasn't at a major university, had no prestigious friends in science and did his work out in the boondocks.

Charles Darwin, in contrast, had taken infinite pains to cultivate a friendly circle of influential biologists. Quick acceptance of his evolution theory followed years of ingratiating letters and a carefully managed campaign to convince the scientific community. In addition to being a great scientist, Darwin was also a consummate diplomat, tactician and public relations man.

Mendel put his faith in the merit of his work, naively believing qualified scientists would surely recognize its worth. He sent a reprint of his paper to the day's leading botanist, Carl Nageli, who was also working with hybrids. Although Nageli then conducted a long correspondence with Mendel, encouraging him to pursue his work, in reality he was no friend to him.

First, he encouraged Mendel to continue testing his theory on hawkweed, instead of pea plants. Hawkweed has a complicated type of inheritance, which was certain to lead to confusing results. And when Nageli published his big book on inheritance and evolution in 1884, there was not a single reference to Mendel.

Ernst Mayr, in 1882, pointed out that Nageli had a complex theory of his own and was one of the few biologists of the time who believed in pure blending inheritance. "To accept Mendel's theory," Mayr concludes, "would have meant, for Nageli, a complete refutation of his own." (*Growth of Biological Theory*, p. 723.)

The cloistered cleric could have used some of the worldliness of the young Thomas Henry Huxley, who made certain to keep one of his papers out of the hands of the great authority in his field, Richard Owen. On March 5, 1852, Huxley wrote a friend:

> You have no idea of the intrigues that go on in this blessed world of science. Science is, I fear, no purer than any other region of human activity; though it should be. Merit alone is very little good; it must be backed by tact and knowledge of the world to do very much.
>
> For instance, I know that the paper I have just sent in is very original and of some importance, and I am equally sure that if it is referred to the judgment of my "particular friend" Professor Owen that it will not be published. He won't be able to say a word against it, but he will pooh-pooh it to a dead certainty.
>
> You will ask with some wonderment, Why? Because for the last twenty years [he] has been regarded as the great authority on these matters, and has had no one to tread on his heels, until at last, I think, he has come to look upon the Natural World as his special preserve, and "no poachers allowed." So I must maneuver a little to get my poor memoir kept out of his hands.

Huxley also used his savvy to "maneuver" his friend Darwin's work from rebel theory to scientific respectability. If Mendel had had a Huxley, his "time" would certainly have come a lot sooner.

Timeline

300 BC. Aristotle obtains twelve duck eggs fertilized at the same time, dissects them on consecutive days, and observes embryogenesis. He proposes a vitalistic force.

1600s. Homunculus Theory. The invention of the microscope leads Antonie van Leeuwenhoek to believe he has seen a "miniature man" in human sperm.

1756. Carl Linnaeus organizes life forms. Single-cell animals reproduce by dividing; multi-cellular animals create an egg which builds a new body.

Early 1800s. Johann von Goethe observes that all life is based on a common form he calls the "Urform."

1859. Charles Darwin proposes natural selection as one possible mechanism for the descent of species of common ancestry.

1874. Wilhelm His discovers that the forms of the body organs may be simulated by the mechanical deformation of inflated tubes and bladders.

1885. Ernst Haeckel states that ontogeny recapitulates phylogeny (the making of the individual from an egg is a fast-forward replay of the history of the species from its single-cell beginning).

1880s. Walther Flemming and others discover chromosomes in the egg.

1880s. August Weismann proposes the "Life Cycle"—the immortal germ line cell dwells in the mortal multi-cellular somatic body, which it constructs in each generation.

1900. Hans Driesch, Wilhelm Roux, and other experimental embryologists find that each of the first eight cells can reproduce the entire body.

1910. Alexander Gurwitsch introduces the concept of the morphogenetic field.

1917. D'Arcy Thompson reports that the forms of the body may be simulated by a variety of mechanical means.

1920. Thomas Hunt Morgan declares that the body form is inherited by the genes rather than the cytoplasm.

1940s. Neo-Darwinism. The dogma called the "Modern Synthesis" declares that evolution is caused by the natural selection of mutations caused by random errors in the transcription of the DNA.

1953. Discovery of the DNA structure by Watson and Crick, believed to be the blueprint of life.

1980s. Stephen Jay Gould and other evolutionary biologists declare that natural selection is irrelevant to the creation of form.

1996. Developmental biologists Scott Gilbert, John Opitz, and Rudolf Raff reaffirm morphogenetic fields as a mechanism of the inheritance of form.

2003. Lynn Margulis announces that the Modern Synthesis is dead.

2008. Sixteen evolutionary biologists meet in Altenberg, Austria, to address the inadequacy of Neo-Darwinism to account for evolution and the inheritance of biological form.

Glossary

Algorithm: a sequence of well-defined instructions for solving a certain problem.
Apterygotes: the wingless predecessors of the insects.
Archenteron: the first body cavity in insects.
Biodiversity: the diverse variations of the phyletic forms.
Blastula: the ball of cells resulting from the cellular subdivision of the egg.
Blastulation: formation of the blastula.
Cambrian: geologic period of the first abundant skeletonized animals.
Cleavage: the subdivision of the egg, or any cell.
Colloid: a liquid/solid state of matter.
Condensation: the disappearance of parts of the sequence of embryological development over generations.
Constrained expansion: the expansion of tissue within a confining space.
Delamination: the separation of contiguous layers.
Developmental mechanics: the theory of embryogenesis by mechanical means, postulated originally by nineteenth-century German scientists and called *Entwicklungsmechanik*.
Embryo: the first form of the organism.
Embryogenesis: the formation of the embryo.
Experimental embryology: an approach to embryology that began in the late nineteenth century in Germany. Novel experiments were designed to manipulate embryos.
Gastrulation: the embryological event where the blastula surface enters the interior to form the embryo.
Gene theory of form: the idea that the body plan of living organisms is encoded in the genes.
Germ layers: the superimposed layers of tissue which constitute the embryo.
Germ line theory: the theory of the separation of the life cycle in the two forms, the germ line and the somatic body.
Germ plasm: the predecessor of the egg.
Heterochrony: the change in the timing of ontogenetic developmental expression.
Homeobox genes: the genes which govern the initiation of the germ band in insects.

Homunculus: the miniature man in the egg or sperm of the preformation theory.

Larva: the intermediate form of some animals between the egg and the adult.

Lepidopteran: relating to butterflies and moths.

Limb primordia: the first structures to appear in limb development.

Mechanical causation: an effect caused by physical forces.

Modern synthesis: the present dogma of biology, which originated in the 1930s, and combined Darwinian natural selection with Mendelian genetics.

Morphogenesis: organic formation.

Morphogenetic field: Briefly, the field is an aspect of a body with a definite boundary that influences the development and growth of a particular organ or organs.

Morphology: having to do with form.

Multi-torus: a torus within a torus, named by Jockusch and Dress (2003).

Mutation: an anomalous form in an individual organism.

Natural selection: the extinction of species that cannot survive the natural environment.

Neo-Darwinism: the modern synthesis (MS).

Ontogeny: development, the course of growth of an individual.

Organogenesis: the formation of the embryonic organs during gastrulation.

Origin of life: the generation of the first living form.

Paradigm shift: term used by Thomas Kuhn, who wrote about the history and philosophy of science, to indicate a situation when change is occurring in a dominant scientific theory.

Phyletic form: the general form of any of the phyla.

Phylogeny: the course of the evolution of the phyla.

Population genetics: the current theory of evolutionary inheritance based on the introduction of a mutation in a population.

Preformation theory: seventeenth-century idea that the body exists in miniature in the egg or sperm.

Primordial germ plasm: August Weismann's name for the predecessor of the egg.

Protoplasm: the fundamental chemical mixture of life, the content of all cells.

Pseudopodal locomotion: the means by which amoeboid cells move.

Reaction diffusion theory: Turing's theory that form is expressed by the chemical gradients which approach from different sides of a tissue.

Recapitulation: the replaying, in an individual's development, of the sequences of its evolution.

Segmentation: axial subdivision of the animal body.

Self-organization: the theory that a given system will become more organized without the help of outside forces. In biology, the phenomenon of self-organization is widely acknowledged to account for the formation of lipid bilayer membranes.

Sol-gel: the intermediate state between two forms of solid-liquid materials.

Torus: a spherical surface with a hole through it, such as a donut.

Urform: the term coined by Goethe for the original form from which all forms are derived.

Urpflanze: Goethe's term for the original flower form.

Wing primordia: the first structures to appear in wing development.

Index

Note that all italic numbers refer to plates.

Permissions

The following have generously given permission to use text and illustrations from the copyrighted works listed below:

Reprinted by permission of the publisher from *Ontogeny and Phylogeny* by Stephen Jay Gould, pp. 2–3; 74–75; 352–358, Cambridge, Mass.: The Belknap Press of Harvard University Press, Copyright © 1977 by the President and Fellows of Harvard College.

From *Developmental Biology,* Eighth Edition Companion Website, Chapter Three, "The Re-discovery of Morphogenetic Fields," by Scott F. Gilbert. Copyright © 2006 by Sinauer Associates, Inc. With kind permission from Sinauer Associates, Inc., Sutherland, MA

From *The Biology of Belief, Unleashing the Power of Consciousness, Matter & Miracles,* by Bruce H. Lipton, PhD. Copyright © 2008 by Mountain of Love Productions. Published by Hay House, Carlsbad, CA. With kind permission from Bruce H. Lipton, Mountain of Love Productions.

From "From Sphere to Torus: A Topological View of the Metazoan Body Plan" by Harald Jockusch and Andreas Dress, in *Bulletin of Mathematical Biology* (2003) Vol. 65: 57–65; DOI: 10.1006/bulm.2002.0319. Copyright © 2002 Society for Mathematical Biology. With kind permission from Springer Science and Business Media.

From "A Gene Regulatory Network Subcircuit Drives a Dynamic Pattern of Gene Expression" by Joel Smith, Christina Theodoris, and Eric H. Davidson, in *Science* (2007) Vol. 318: 794–797; DOI: 10.1126/science.1146524. Copyright © 2007 by AAAS. Reprinted with permission from AAAS.

From "Topological Topological Patterns in Metazoan Evolution and Development" by Valeria Isaeva, Eugene Presnov, and Alexey Chernyshev, in *Bulletin of Mathematical Biology* (2006) Vol. 68: 2053–2067; DOI 10.1007/s11538-006-9063-2. Copyright © 2006 Society for Mathematical Biology. With kind permission from Springer Science and Business Media.

From "Broken Symmetries and Biological Patterns" by Ian Stewart, in *On Growth, Form and Computers* (Sanjeev Kumar, Peter J. Bentley, eds.), Chapter Ten, pp. 181–183. Published by Elsevier Academic Press, Amsterdam. Copyright © 2003 by Elsevier Ltd. With kind permission from Elsevier Ltd.

Acknowledgments

I would like to gratefully acknowledge the many contributors who helped with this project over the past ten years. Without their dedication and zeal this book would not have been possible:

Sama Ali, Ryan Beckwith, Rob DeSalle, Murray Gell-Mann, Eliot Goldfinger, Stephen Jay Gould, Stephen Halker, Kathy Hall, Melissa Ludwig, Richard Milner, Peter Sheesley, Sammy Shin, Carole Syrota, Ryan Toth, Kate Waring, and with special thanks to Janusz Kusyk and Larimore Hampton.

Human Blueprint Laboratory

From left: Peter Sheesley, Kathy Hall, Carole Syrota, Sammy Shin, Ryan Toth, Stephen Halker, Stuart Pivar, and Melissa Ludwig.

About the Author

Stuart Pivar was born in 1930 in Brooklyn, New York. He attended Brooklyn Technical High School, where he studied chemistry and engineering, and received a BA in chemistry from Hofstra University in 1951.

Pivar went on to found Chem-Tainer Industries in 1959 and holds a number of patents on processes and machinery for plastic molding. The company produces industrial liquid containment vessels at a number of manufacturing facilities throughout the U.S.

After a quarter-century of collecting art and being fully immersed in the artistic milieu, in 1980 Pivar started the New York Academy of Art, an accredited graduate art school grounded in the academic tradition.

In the mid-1990s Pivar's attention shifted exclusively to scientific pursuits, founding Protease Inhibitor Labs at SUNY Stony Brook (for AIDS research, 1995–1999), Regenerative Medicine Labs (investigating therapeutic cloning, 1998–2002), and the Human Blueprint Laboratory (investigating origins of form, 1996 to the present).

Pivar is a long-time resident of New York City.